Advances in Intelligent Systems and Computing

Volume 1006

The series "Advances in Intelligent Systems and Computing" contains publications on theory, applications, and design methods of Intelligent Systems and Intelligent Computing. Virtually all disciplines such as engineering, natural sciences, computer and information science, ICT, economics, business, e-commerce, environment, healthcare, life science are covered. The list of topics spans all the areas of modern intelligent systems and computing such as: computational intelligence, soft computing including neural networks, fuzzy systems, evolutionary computing and the fusion of these paradigms, social intelligence, ambient intelligence, computational neuroscience, artificial life, virtual worlds and society, cognitive science and systems, Perception and Vision, DNA and immune based systems, self-organizing and adaptive systems, e-Learning and teaching, human-centered and human-centric computing, recommender systems, intelligent control, robotics and mechatronics including human-machine teaming, knowledge-based paradigms, learning paradigms, machine ethics, intelligent data analysis, knowledge management, intelligent agents, intelligent decision making and support, intelligent network security, trust management, interactive entertainment, Web intelligence and multimedia.

The publications within "Advances in Intelligent Systems and Computing" are primarily proceedings of important conferences, symposia and congresses. They cover significant recent developments in the field, both of a foundational and applicable character. An important characteristic feature of the series is the short publication time and world-wide distribution. This permits a rapid and broad dissemination of research results.

**** Indexing: The books of this series are submitted to ISI Proceedings, EI-Compendex, DBLP, SCOPUS, Google Scholar and Springerlink ****

More information about this series at http://www.springer.com/series/11156

Paulo Novais · Jaime Lloret ·
Pablo Chamoso · Davide Carneiro ·
Elena Navarro · Sigeru Omatu
Editors

Ambient Intelligence – Software and Applications –,10th International Symposium on Ambient Intelligence

Editors
Paulo Novais
Departamento de Informática
Universidade do Minho
Braga, Portugal

Jaime Lloret
Department of Communications
Polytechnic University of Valencia
Valencia, Valencia, Spain

Pablo Chamoso
IoT European Digital Innovation Hub
University of Salamanca
Salamanca, Salamanca, Spain

Davide Carneiro
Departamento de Informática
Universidade do Minho
Braga, Portugal

Elena Navarro
Escuela de Ingenieros Industriales de Albacete, Departamento de Sistemas Informáticos
Universidad de Castilla-La Mancha
Albacete, Spain

Sigeru Omatu
Faculty of Robotics and Design Engineering
Osaka Institute of Technology
Osaka, Japan

ISSN 2194-5357 ISSN 2194-5365 (electronic)
Advances in Intelligent Systems and Computing
ISBN 978-3-030-24096-7 ISBN 978-3-030-24097-4 (eBook)
https://doi.org/10.1007/978-3-030-24097-4

This Springer imprint is published by the registered company Springer Nature Switzerland AG
The registered company address is: Gewerbestrasse 11, 6330 Cham, Switzerland

Preface

Ambient intelligence (AmI) is a paradigm emerging from artificial intelligence, where computers are used as proactive tools assisting people with their day-to-day activities, making everyone's life more comfortable. Another main concern of AmI originates from the human–computer interaction domain and focuses on offering ways to interact with systems in a more natural way by means user-friendly interfaces. This field is evolving quickly as can be witnessed by the emerging natural language and gesture-based types of interaction.

The inclusion of computational power and communication technologies in everyday objects is growing, and their embedding into our environments should be as invisible as possible. In order for AmI to be successful, human interaction with computing power and embedded systems in the surroundings should be smooth and happen without people actually noticing it. The only awareness people should have arisen from AmI: more safety, comfort, and wellbeing, emerging in a natural and inherent way. ISAmI is the International Symposium on Ambient Intelligence, aiming to bring together researchers from various disciplines that constitute the scientific field of AmI to present and discuss the latest results, new ideas, projects, and lessons learned. Brand new ideas will be greatly appreciated as well as relevant revisions and actualizations of previously presented work, project summaries, and PhD thesis.

This year's technical program will present both high quality and diversity, with contributions in well-established and evolving areas of research. Specifically, 45 papers were submitted by authors from over 10 different countries (Brazil, Italy, Japan, Morocco, Portugal, Spain, Tunisia, or United States, among others), representing a truly "wide area network" of research activity. The ISAmI technical program has selected 22 papers, and as in past editions, it will be special issues in JCR-ranked journals such as information fusion, neurocomputing, sensors, processes, and electronics. Moreover, ISAmI'19 workshops have been a very useful tool in order to complement the regular program with new or emerging topics of particular interest to the participating community.

This symposium is organized by the Universidade do Minho, Technical University of Valencia, Hiroshima University, and University of Salamanca. The present edition was held in Avila, Spain, from June 26–28, 2019. We thank the sponsors IEEE Systems Man and Cybernetics Society Spain Section Chapter and the IEEE Spain Section (Technical Co-Sponsor), IBM, Indra, Viewnext, Global Exchange, AEPIA, APPIA and AIR institute, as well as the support of the Regional Government de Castilla y León (Spain) with the project *"Desarrollo de Capacidades Tecnológicas en torno a la Aplicación Industrial de Internet de las Cosas (IOTEC)"* (Id. 0123_IOTEC_3_E- Project co-financed with FEDER funds, Interreg España-Portugal (PocTep)), and finally, the local organization members and the program committee members for their hard work, which was essential for the success of DCAI'19.

June 2019

Paulo Novais
Jaime Lloret
Pablo Chamoso
Davide Carneiro
Elena Navarro
Sigeru Omatu

Organization

General Chairs

Paulo Novais	Universidade do Minho, Portugal
Jaime Lloret	Universitat Politecnica de Valencia, Spain

Organizing Committee Chairs

Sigeru Omatu	Hiroshima University, Japan
Pablo Chamoso	University of Salamanca, Spain
Davide Carneiro	Intelligent Systems Lab, Universidade do Minho, Portugal

Program Committee

Ana Almeida	ISEP-IPP, Portugal
Ana Alves	Centre for Informatics and Systems, University of Coimbra, Portugal
Ricardo Anacleto	ISEP, Portugal
Cesar Analide	University of Minho, Portugal
Cecilio Angulo	Universitat Politècnica de Catalunya, Spain
Lars Braubach	University of Hamburg, Germany
Maria-Pilar Cáceres-Reche	Department of Didactic and School Organization, Faculty of Sciences of Education, Spain
Valérie Camps	University of Toulouse, IRIT, France
Javier Carbo	University Carlos III of Madrid, Spain
Gonçalo Cardeal	Universidade de Lisboa, Portugal
Davide Carneiro	Polytechnic Institute of Porto, Portugal
Joao Carneiro	ISEP/GECAD, Portugal
Fabio Cassano	Università degli Studi di Bari Aldo Moro, Italy
José Antonio Castellanos Garzón	University of Salamanca, Spain

Jose Carlos Castillo Montoya	Universidad Carlos III de Madrid, Spain
Alvaro Castro-Gonzalez	Universidad Carlos III de Madrid, Spain
João P. S. Catalão	University of Porto, Portugal
Silvio Cesar Cazella	UFCSPA, Brazil
Pablo Chamoso	University of Salamanca, Spain
Stefano Chessa	Department of Computer Science, University of Pisa, Italy
Stéphanie Combettes	IRIT, University of Toulouse, France
Luís Conceição	GECAD, Research Group on Intelligent Engineering and Computing for Advanced Innovation and Development, Portugal
Phan Cong-Vinh	Nguyen Tat Thanh University, Vietnam
Ricardo Costa	ESTG-IPP, Portugal
Rémy Courdier	LIM, Université de la Réunion, Reunión
Fernando De La Prieta	University of Salamanca, Réunion
Patricio Domingues	ESTG Leiria, Portugal
John Dowell	University College London, UK
Dalila Duraes	Department of Artificial Intelligence, Technical University of Madrid, Madrid, Spain
Luiz Faria	Knowledge Engineering and Decision Support Research (GECAD), Institute of Engineering, Polytechnic of Porto, Porto, Portugal
Florentino Fdez-Riverola	University of Vigo, Spain
Marta Fernandes	GECAD, Research Group on Intelligent Engineering and Computing for Advanced Innovation and Development, Polytechnic of Porto, Portugal
Bruno Fernandes	University of Minho, Portugal
Antonio Fernández-Caballero	Universidad de Castilla-La Mancha, Spain
João Ferreira	ISCTE, Portugal
Lino Figueiredo	ISEP, Portugal
Adina Magda Florea	University Politehnica of Bucharest, AI-MAS Laboratory, Romania
Daniela Fogli	Università di Brescia, Italy
Celestino Goncalves	Instituto Politecnico da Guarda, Portugal
Sérgio Gonçalves	University of Minho, Portugal
Alfonso González Briones	BISITE Research Group, Spain
David Griol	Universidad Carlos III de Madrid, Spain
Junzhong Gu	East China Normal University, China
Esteban Guerrero	Umeå University, Sweden
Hans W. Guesgen	Massey University, New Zealand
Javier Jaen	Universitat Politècnica de València, Spain
Jean-Paul Jamont	LCIS, Université de Grenoble, France
Vicente Julian	Universitat Politècnica de València, Spain

Alberto Rivas	BISITE Research Group, University of Salamanca, Spain
Sara Rodríguez	University of Salamanca, Spain
Teresa Romão	Faculdade de Ciências e Tecnologia/Universidade NOVA de Lisboa (FCT/UNL), Portugal
Albert Ali Salah	Bogazici University, Turkey
Altino Sampaio	Instituto Politécnico do Porto, Escola Superior de Tecnologia e Gestão de Felgueiras, Portugal
Manuel Filipe Santos	University of Minho, Portugal
Enzo Pasquale Scilingo	University of Pisa, Italy
Fernando Silva	Department of Informatics Engineering; School of Technology and Management; Polytechnic Institute of Leiria, Portugal
Fábio Silva	University of Minho, Portugal
S. Shyam Sundar	Penn State University and Sungkyunkwan University, USA/Korea
Radu-Daniel Vatavu	University Stefan cel Mare of Suceava, Romania
Lawrence Wai-Choong Wong	National University of Singapore, Singapore
Ansar-Ul-Haque Yasar	Universiteit Hasselt, IMOB, Belgium

Organization Committee

Juan Manuel Corchado Rodríguez	University of Salamanca, Spain, and AIR institute, Spain
Pablo Chamoso Santos	University of Salamanca, Spain
Sara Rodríguez González	University of Salamanca, Spain
Fernando De la Prieta	University of Salamanca, Spain
Sonsoles Pérez Gómez	University of Salamanca, Spain
Benjamín Arias Pérez	University of Salamanca, Spain
Javier Prieto Tejedor	University of Salamanca, Spain, and AIR institute, Spain
Amin Shokri Gazafroudi	University of Salamanca, Spain
Alfonso González Briones	University of Salamanca, Spain, and AIR institute, Spain
José Antonio Castellanos	University of Salamanca, Spain
Yeray Mezquita Martín	University of Salamanca, Spain
Enrique Goyenechea	University of Salamanca, Spain
Javier J. Martín Limorti	University of Salamanca, Spain
Alberto Rivas Camacho	University of Salamanca, Spain
Ines Sitton Candanedo	University of Salamanca, Spain
Daniel López Sánchez	University of Salamanca, Spain
Elena Hernández Nieves	University of Salamanca, Spain
Beatriz Bellido	University of Salamanca, Spain

María Alonso	University of Salamanca, Spain
Diego Valdeolmillos	University of Salamanca, Spain, and AIR institute, Spain
Roberto Casado Vara	University of Salamanca, Spain
Sergio Marquez	University of Salamanca, Spain
Guillermo Hernández González	University of Salamanca, Spain
Mehmet Ozturk	University of Salamanca, Spain
Luis Carlos Martínez de Iturrate	University of Salamanca, Spain, and AIR Institute, Spain
Ricardo S. Alonso Rincón	University of Salamanca, Spain
Javier Parra	University of Salamanca, Spain
Niloufar Shoeibi	University of Salamanca, Spain
Zakieh Alizadeh-Sani	University of Salamanca, Spain
Jesús Ángel Román Gallego	University of Salamanca, Spain
Angélica González Arrieta	University of Salamanca, Spain
José Rafael García-Bermejo Giner	University of Salamanca, Spain
Pastora Vega Cruz	University of Salamanca, Spain
Mario Sutil	University of Salamanca, Spain
Belén Pérez Lancho	University of Salamanca, Spain
Angel Luis Sánchez Lázaro	University of Salamanca, Spain

Contents

10th International Symposium on Ambient Intelligence

Computer-Aided Hepatocarcinoma Diagnosis Using Multimodal Deep Learning

Alan Baronio Menegotto(✉), Carla Diniz Lopes Becker, and Silvio Cesar Cazella

Universidade Federal de Ciencias da Saude de Porto Alegre, Rua Sarmento Leite, 245, Porto Alegre, Rio Grande do Sul, Brazil
{alanb,carladiniz,silvioc}@ufcspa.edu.br

Abstract. Liver cancer was the fourth most deadly cancer in 2018 worldwide. Among liver cancers, hepatocarcinoma is the most prevalent cancer type. Diagnostic protocols are complex and suggest variations based on the patient's context and the use of multiple data modalities. This paper briefly describes the steps involved in the development of a hepatocarcinoma computer-aided diagnosis using a multimodal deep learning approach with imaging and tabular data fusion. Data acquisition, preprocessing steps, architectural design decisions and possible use cases for the described architecture are discussed based on the partial results achieved on this ongoing research.

Keywords: e-health · Hepatocarcinoma · Computer-aided diagnosis · Multimodal deep learning

1 Introduction

According to the World Health Organization estimates, in 2018 approximately 18 million people worldwide were affected by some type of cancer and of these people, 9,6 million people died of the disease. The types of cancer with the highest mortality rate in 2018 were: Lung (1,761,007 deaths), Colo-rectal (880,792 deaths), Stomach (782,685 deaths), Liver (781,631 deaths) and Breast (626,679 deaths) [1]. Despite recent advances in cancer diagnosis and treatment, estimates from different institutes point to a substantial rise in cancer cases in the coming decades caused mainly by factors such as environment, stressful routines and ageing population [2–4].

Hepatocarcinoma is the most common liver cancer and has pathognomonic symptoms [5]. The diagnostic protocol recommended by reference institutes suggests the use of multiple inputs such as imaging, clinical attributes and behaviour information depending on each patient's context [6,7]. The subtleties involved in diagnostic demands a high level of attention and at the end of a long and busy workday the physician's accuracy can be impaired. An unobserved single detail in this context can put the patient's life at risk.

P. Novais et al. (Eds.): ISAmI 2019, AISC 1006, pp. 3–10, 2020.
https://doi.org/10.1007/978-3-030-24097-4_1

The increasing number of cancer cases and long workdays will impose physicians to use tools that assess a quick and precise diagnostic. Computer-aided diagnosis is a broad research area which aims to develop and evolve these tools. A computer system can acquire the intelligence needed for tasks such as the disease's diagnosis and life expectancy prediction through artificial intelligence approaches [8].

The popularization of Electronic Health Record (EHR) systems that store information entered manually by health professionals and integrated data from multiple sources such as patient monitoring solutions, Laboratory Information System (LIS) and Picture Archiving and Communication Systems (PACS) present a valuable opportunity for the development and use of multimodal computer-aided diagnosis systems. These systems could process and correlate distinct modalities extracted from EHR and increase the precision of the suggested predictions. A precise and trusty Computer-Aided Diagnosis (CAD) system can relieve the physician work burden in diagnostic process and help physicians to keep the same diagnostic performance all day long.

The paper's main goal is to describe the development of a multimodal deep machine learning architecture for hepatocarcinoma diagnostic assistance that uses computed tomography images, laboratory test results, anthropometric and sociodemographic data as input. The paper is structured as follows: Sect. 2 shows related researches found in the literature. Section 3 describes the data acquisition and preprocessing stage. Section 4 discusses the multimodal deep learning architecture design used and explores possible applications that can be implemented using the described architecture and Sect. 5 contains final considerations related to the discussed subject.

2 State of the Art

Artificial intelligence techniques have already been used for a long time to diagnose liver-related pathologies. However, it was only with hardware's evolution and significant advances in techniques such as Deep Learning [9] that the precision of the diagnosis performed exclusively by computers began to be comparable with the accuracy of medical specialists [10]. Monomodal and ensemble machine learning are the most popular approaches found in the literature for liver-related pathologies CAD systems [11,12].

Using more than one data modality as input for machine learning algorithms is known as multimodal machine learning [13]. Multimodal machine learning approaches to diagnosing liver-related pathologies is a hot research topic. Ali et al. [14] uses imaging and clinical data with Support Vector Machine (SVM). Wang et al. [15] uses Random Forests (RF) to process clinical and genetic data. Deep multimodal machine learning using extracted features of imaging exams [16], different planes [17] or dimensions [18] of the same imaging study are also found in the literature.

The approach proposed in this paper can spot some light on how to increase accuracy in hepatocarcinoma diagnosing using deep neural networks. Moreover, processing just one image alongside tabular data is cheaper from a computational point of view than the deep multimodal machine learning approaches mentioned above. However, the proposed approach in practice can be used only by healthcare institutions that have all the data modalities needed as input stored in an EHR.

3 Data Acquisitions and Preprocessing

Data quality used as input in deep learning algorithms is one of the key factors for its success [9]. Image data should be crystal clear. Tabular data should be normalized, without missing information and has a homogeneous representation. These requirements represent a utopia because in practice data obtained from the real world usually does not fit these conditions. Therefore a data preprocessing phase is usually needed in deep learning implementations.

In this study, image and tabular data were obtained from different databases of the US National Cancer Institute's Genomic Data Commons (GDC) platform: TCGA-LIHC [19] for positive hepatocarcinoma cases and TCGA-STAD [20], TCGA-KIRP [21] and CPTAC-PDA [22] for negative hepatocarcinoma cases. From these datasets, abdominal CT studies with an axial view were extracted. After that, the slices that show the liver were extracted to compose the database described in Table 1:

Table 1. Image dataset summary.

Source database	Size	Slices	Patients	Studies	Class
TCGA-LIHC	10.2 GB	20,792	73	402	Positive
TCGA-KIRP	1.82 GB	1,968	22	48	Negative
TCGA-STAD	6.76 GB	13,741	46	184	Negative
CPTAC-PDA	5.08 GB	10,331	33	132	Negative

CT images are very noisy due to the capture process. In addition to the required unit conversions for Digital Imaging and Communications in Medicine (DICOM) standard image processing, another step is also required to improve the sharpness of the image and decrease its size without loss of relevant information. To enhance and compress the images, a three-step process was performed in each slice: (1) The enhancement of the hepatobiliary structure was performed with Contrast Limited Adaptive Histogram Equalization (CLAHE) [23], an algorithm which comes successfully being used for medical image enhancement [24]. (2) Noise removal was performed with the Total Variation Denoising algorithm written by Chambolle [25], which demonstrated excellent visual results when compared to other algorithms such as Wavelet and Bilateral Denoising. (3) Image lossless compression is performed at the time of PNG generation, one of the possible file formats for convolutional neural networks image input. Figure 1 shows an example of the image preprocessing step result with a red arrow that indicates where is located the liver. After preprocessing, the images were divided into training, validation and test sets in 70/20/10 ratio.

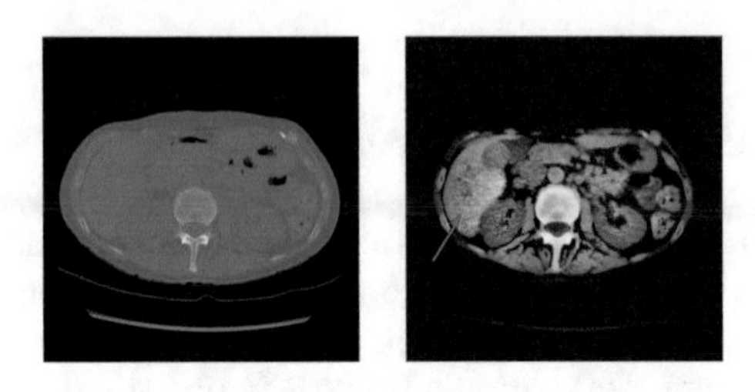

Fig. 1. Image before and after preprocessing

Tabular patient data containing laboratory tests results, anthropometric and sociodemographic information are also provided by the GDC platform. The relationship between imaging studies and tabular data is made by the patient identifier in the GDC platform.

The information provided in the tabular dataset was processed and grouped to keep only relevant attributes for hepatocarcinoma diagnosis. The attribute selection to compose the tabular dataset was based on the European Association for the Study of Liver (EASL) diagnostic and treatment protocol [7]. After attributes selection, missing data imputation was performed using techniques such as Artificial Neural Networks, Random Forests, Pseudo-Random Numbers Generated using a normal distribution and spline interpolation. At the end of this step, all patients have the following attributes:

- Anthropometric and Sociodemographic: Source, Patient, Gender, Age at Diagnosis, Height, Weight, Race e Ethnicity
- Clinical: Other Malignancy, Family History Cancer Indicator, Family History Cancer Number Relatives, Alcohol Consumption, Hemochromatosis, Hepatitis, Non-Alcoholic Fatty Liver Disease, Other Liver Disease
- Laboratory Tests Results: Alpha-Fetoprotein, Platelets, Prothrombin Time, Albumin, Bilirubin, Creatinine.

4 Multimodal Deep Learning Architecture

The input of multiple data modalities to machine learning algorithms is performed through data fusion techniques [26]. Data fusion is classified based on the stage where the fusion occurs: (1) Early Fusion occurs when the data is merged before the input layer of the machine learning algorithm; (2) Intermediate Fusion, also known as joint fusion, occurs in the network intermediary layers when all data modalities has the same representation format; (3) Late Fusion occurs when each data modality is processed end-to-end separately and then combined into one representation for further processing [27].

Early fusion imaging and tabular data could confuse the Deep Convolutional Neural Network (DCNN) in imaging feature extraction task. Late fusion would require more steps in network architecture. So, the Intermediate Fusion approach was chosen for the proposed multimodal deep learning architecture: a DCNN is used to extract image features and before classification layer, tabular data of each patient is concatenated with the flattened output of the image feature extraction.

DCNN training for complex images classification is a computationally costly task. One way to avoid this cost is fine-tuning a pre-trained state-of-the-art network with new classes of images. This technique is called knowledge transfer [28]. By using knowledge transfer, the network already has the necessary knowledge to extract the basic features of an image (edges, contours, textures, lines, etc ...) and the training serves just to adapt the final layers with the new inputs and outputs.

Inception V3 [29] is a DCNN created by Google that achieves 5.64% top-5 error on the validation set of the whole image ImageNet Large Scale Visual Recognition Challenge (ILSVRC) 2012 classification task and placed 2nd on ILSVRC 2015 contest. Inception V3 is considered a state-of-the-art network and has already been used to classify medical images using knowledge transfer achieving great results [8]. Therefore, Inception V3 is the state-of-the-art DCNN chosen for the development of the hepatocarcinoma CAD system. Figure 2 was based on [30] and shows the developed architecture. The architecture was implemented tweaking the Inception V3 implementation that comes out-of-the-box in Keras [31], a popular python deep learning framework. Amazon EC2 infrastructure was used for the fine-tuning task as it has the best cost-benefit ratio in General Processing Unit (GPU) computing.

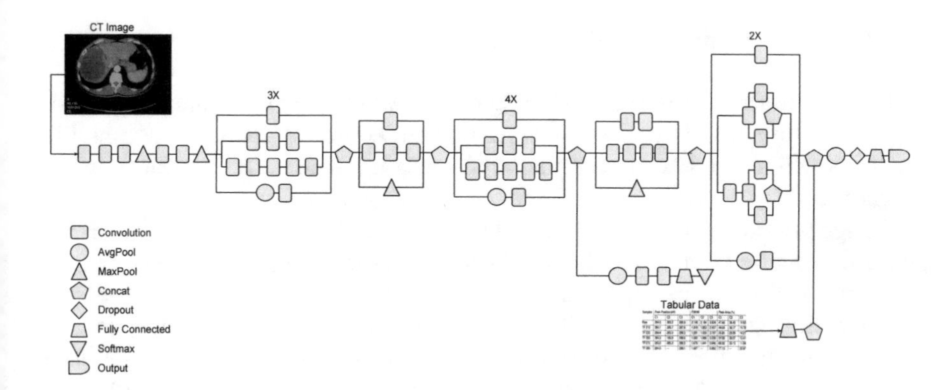

Fig. 2. Multimodal deep learning architecture

Usage of knowledge acquired in training time by the multimodal deep learning architecture allows the development of CAD applications for hepatocarcinoma. These applications can be designed for interactive or non-interactive usage: web, desktop or mobile applications can be developed as well as batch processes. There are three ways to encapsulate and reuse this knowledge:

1. Reuse the saved DCNN weights in the Keras framework locally through a shell script or a program developed with the Python programming language.
2. Create a Representational State Transfer (REST) endpoint that receives images and tabular data as input for the developed DCNN and returns a hepatocarcinoma diagnosis. The REST approach becomes interesting because it allows the DCNN remote use by different health systems transparently to end-users;
3. Create a web CAD application that allows the upload of images and tabular data as input and uses the developed DCNN to return a hepatocarcinoma diagnosis. This approach allows the interactive use of DCNN through web browsers by end-users and also through HTTP calls that can be done by third-party systems in a non-interactive way.

Hepatocarcinoma is a pathognomonic disease and often the affected person does not know about its condition. Early cancer diagnosis is related to shorter treatments and increased surveillance rates [32]. It is possible to develop a batch application that consumes the developed DCNN in one of the ways described above and routinely scan healthcare institutions EHR records searching for possible positive cases of hepatocellular carcinoma that were not diagnosed yet. When a positive case is found, the application can trigger actions like sending alerts for physicians, request further examination or making a doctor appointment, actions that can be considered as the treatment begin. It's a simple idea that can save money and increase surveillance rates just by anticipating diagnosis and treatment's begin.

5 Final Considerations

This article describes part of an ongoing project whose goal is to quantify the gain in accuracy of a multimodal deep learning architecture compared to a unimodal DCNN architecture that works just with imaging exams for hepatocarcinoma diagnosis.

Image preprocessing steps used in this article can lead to better results in classification accuracy even for other artificial intelligence projects that use abdominal axial CT slices as input. The architecture described uses the knowledge acquired during the training phase conducted by Google in the Inception V3 network and fine-tune this knowledge to the hepatocarcinoma diagnosis using multiple data modalities following the best practices found in the literature.

Hepatocarcinoma is a silent disease that can lead to death if diagnosed in advanced stages. The creation of a CAD system that can be used in an interactive and non-interactive way is of great value for healthcare institutions that can proactively detect possible cases of hepatocarcinoma with the information already stored in their EHR systems and start the patient's treatment as soon as possible, increasing survival probabilities.

References

1. International Agency for Research on Cancer: Global Cancer Observatory - Cancer Today (2018). http://gco.iarc.fr/today/online-analysis-table?v=2018&mode=cancer&mode_population=continents&population=900&populations=900&key=asr&sex=0&cancer=39&type=1&statistic=5&prevalence=0&population_group=0&ages_group%5B%5D=0&ages_group%5B%5D=17&nb_items=5&gr
2. Stewart, B., Wild, C.: World Healt Report. International Agency for Research on Cancer, Lyon (2014)
3. Rahib, L., Smith, B.D., Aizenberg, R., Rosenzweig, A.B., Fleshman, J.M., Matrisian, L.M.: Projecting cancer incidence and deaths to 2030: the unexpected burden of thyroid, liver, and pancreas cancers in the united states. Cancer Res. **74**, 2913–2921 (2014)
4. Cancer Research UK: Worldwide cancer incidence statistics—Cancer Research UK (2014). http://www.cancerresearchuk.org/health-professional/cancer-statistics/worldwide-cancer/incidence#heading-Five
5. Attwa, M.H., El-Etreby, S.A.: Guide for diagnosis and treatment of hepatocellular carcinoma. World J. Hepatol. **7**, 1632–1651 (2015)
6. Bruix, J., Sherman, M.: Management of hepatocellular carcinoma: an update. Hepatol. **53**, 1020–1022 (2011)
7. Galle, P.R., Forner, A., Llovet, J.M., Mazzaferro, V., Piscaglia, F., Raoul, J.L., et al.: EASL clinical practice guidelines: management of hepatocellular carcinoma. J. Hepatol. **69**, 182–236 (2018)
8. Esteva, A., Kuprel, B., Novoa, R.A., Ko, J., Swetter, S.M., Blau, H.M., et al.: Dermatologist-level classification of skin cancer with deep neural networks. Nature **542**, 115–118 (2017)
9. Litjens, G., Kooi, T., Bejnordi, B.E., Setio, A.A.A., Ciompi, F., Ghafoorian, M., et al.: A survey on deep learning in medical image analysis. Med. Image Anal. **42**, 60–88 (2017)
10. Esses, S.J., Lu, X., Zhao, T., Shanbhogue, K., Dane, B., Bruno, M., et al.: Automated image quality evaluation of T2-weighted liver MRI utilizing deep learning architecture. J. Magn. Reson. Imaging **47**, 723–728 (2018)
11. Ben-Cohen, A., Klang, E., Diamant, I., Rozendorn, N., Raskin, S.P., Konen, E., et al.: CT Image-based decision support system for categorization of liver metastases into primary cancer sites: initial results. Acad. Radiol. **24**, 1501–1509 (2017)
12. Sathurthi, S., Saruladha, K.: Prediction of liver cancer using random forest ensemble. Int. J. Pure Appl. Math. **116**(21), 267–273 (2017)
13. Ngiam, J., Khosla, A., Kim, M., Nam, J., Lee, H., Ng, A.Y.: Multimodal deep learning. In: Proceedings of the 28th International Conference on Machine Learning (ICML-11), pp. 689–696 (2011)
14. Ali, L., Khelil, K., Wajid, S.K., Hussain, Z.U., Shah, M.A., Howard, A., et al.: Machine learning based computer-aided diagnosis of liver tumours. In: 2017 IEEE 16th International Conference on Cognitive Informatics & Cognitive Computing (ICCI*CC), pp. 139–145. IEEE (2017). http://ieeexplore.ieee.org/document/8109742/
15. Wang, J., Jain, S., Chen, D., Song, W., Hu, C.T., Su, Y.H.: Development and evaluation of novel statistical methods in urine biomarker-based hepatocellular carcinoma screening. Sci. Rep. **8**(1), 3799 (2018). http://www.nature.com/articles/s41598-018-21922-9

16. Dou, T., Zhang, L., Zheng, H., Zhou, W.: Local and non-local deep feature fusion for malignancy characterization of hepatocellular carcinoma, pp. 472–479. Springer, Cham (2018). http://link.springer.com/10.1007/978-3-030-00937-3_54
17. Wang, Q., Zhang, L., Xie, Y., Zheng, H., Zhou, W.: Malignancy characterization of hepatocellular carcinoma using hybrid texture and deep features. In: 2017 IEEE International Conference on Image Processing (ICIP), pp. 4162–4166. IEEE (2017). http://ieeexplore.ieee.org/document/8297066/
18. Dou, T., Zhou, W.: 2D and 3D convolutional neural network fusion for predicting the histological grade of hepatocellular carcinoma. In: 2018 24th International Conference on Pattern Recognition (ICPR), pp. 3832–3837. IEEE (2018). https://ieeexplore.ieee.org/document/8545806/
19. Erickson, B.J., Kirk, S., Lee, Y., Bathe, O., Kearns, M., Gerdes, C., et al.: TCGA-LIHC - The Cancer Imaging Archive (TCIA) Public Access - Cancer Imaging Archive Wiki (2016). https://wiki.cancerimagingarchive.net/display/Public/TCGA-LIHC#49e04d416a274e2c9a1218c4350512e9
20. Lucchesi, F.R., Aredes, N.D.: Radiology data from the cancer genome atlas stomach adenocarcinoma [TCGA-STAD] collection (2016). https://wiki.cancerimagingarchive.net/display/Public/TCGA-STAD
21. Linehan, M., Gautam, R., Kirk, S., Lee, Y., Roche, C., Bonaccio, E., et al.: Radiology data from the cancer genome atlas cervical kidney renal papillary cell carcinoma [KIRP] collection (2016). https://wiki.cancerimagingarchive.net/display/Public/TCGA-KIRP
22. National Cancer Institute Clinical Proteomic Tumor Analysis Consortium: Radiology data from the clinical proteomic tumor analysis consortium pancreatic ductal adenocarcinoma [CPTAC-PDA] collection (2018). https://wiki.cancerimagingarchive.net/display/Public/cptac-pda
23. Pizer, S.M., Philip Amburn, E., Austin, J.D., Cromartie, R., Geselowitz, A., Greer, T., et al.: Adaptive histogram equalization and its variations. Comput. Vis. Graph. Image Process. **39**, 355–368 (1987)
24. Reza, A.M.: Realization of the contrast limited adaptive histogram equalization (CLAHE) for real-time image enhancement. J. VLSI Sig. Process. Syst. Sig. Image Video Technol. **38**, 35–44 (2004)
25. Chambolle, A.: An algorithm for total variation minimization and applications. J. Math. Imaging Vis. **20**, 89–97 (2004)
26. Lahat, D., Adali, T., Jutten, C.: Multimodal data fusion: an overview of methods, challenges, and prospects. Proc. IEEE **103**, 1449–1477 (2015)
27. Ramachandram, D., Taylor, G.W.: Deep multimodal learning: a survey on recent advances and trends. IEEE Sig. Process. Mag. **34**, 96–108 (2017)
28. Tajbakhsh, N., Shin, J.Y., Gurudu, S.R., Hurst, R.T., Kendall, C.B., Gotway, M.B., et al.: Convolutional neural networks for medical image analysis: full training or fine tuning? IEEE Trans. Med. Imaging **35**, 1299–1312 (2016)
29. Szegedy, C., Vanhoucke, V., Ioffe, S., Shlens, J.: Rethinking the inception architecture for computer vision (2016)
30. Shlens, J.: Google AI Blog (2016). https://ai.googleblog.com/2016/03/train-your-own-image-classifier-with.html
31. Chollet, F.: Keras (2015). https://keras.io
32. Smith, R.A., Andrews, K.S., Brooks, D., Fedewa, S.A., Manassaram-Baptiste, D., Saslow, D., et al.: Cancer screening in the United States, 2018: a review of current American cancer society guidelines and current issues in cancer screening. CA Cancer J. Clin. **68**, 297–316 (2018)

Multi-Agent System and Classification Algorithms Applied for eHealth in Order to Support the Referral of Post-operative Patients

Tibério C. J. Loureiro(✉), Afonso B. L. Neto(✉), Francisco A. A. Rocha(✉), Francisca A. R. Aguiar(✉), and Marcial P. Fernandez(✉)

Universidade Estadual do Ceará (UECE), Fortaleza, Ceara, Brazil
tiberiocj@gmail.com, afonsoblneto@gmail.com, alexandresobral2004@gmail.com, alannyrocha2009@hotmail.com, marcial@larces.uece.br

Abstract. The need to perform accurate and timely diagnoses in cases involving patients in a post-operative situation is one of the challenges involved in the area of health care. According to studies in some countries, patients are concerned about spending more time in the hospital after surgery. In this sense, we tried to verify how the implementation of solutions that use IT devices and techniques could improve this process of diagnosis. Considering those, this article proposes a multi-agent system architecture that uses, among other techniques, IoT devices, machine learning algorithms and the XMPP protocol with the purpose of determining the best referral to post-operative patients based on medical information. The results obtained showed an accuracy of almost 90% in the evaluated cases, evidencing the possibility of the use and evolution of the IT solution that was developed.

Keywords: Multi-agent system · Health · Post-operative · IoT · Machine Learning

1 Introduction

According to Yu [15], almost 70% to 80% of North American patients expressed a preference for a day surgery. Recent research has shown health as the main concern of people, being overcome, in some situations or countries, only by economic survival issues such as unemployment and insufficient income [17]. This relevance makes health policy, actions and services more important in the contemporary societies agenda.

Increasingly, information provides a new point of view for the interpretation of events or phenomena, which provide visibility and previously invisible meanings [7]. In this way, health companies, clinics and hospitals are developing strategies for collect and analyze data to optimize patients' care and facilitate the identification of infection risk patterns.

P. Novais et al. (Eds.): ISAmI 2019, AISC 1006, pp. 11–18, 2020.
https://doi.org/10.1007/978-3-030-24097-4_2

Among these strategies, we highlight the applications of IoTs devices that can potentially offer new tools for healthcare systems, including hospital emergency cases treatment, long-term treatment and community-based treatment [6].

Therefore, since these technologies stand out in the health sector as an innovative strategy that qualifies the service, it contributes to risk reduction and early diagnosis of health problems. This article presents a new solution for post-operative patient diagnosis. It proposes using IoT devices and Machine Learning techniques to assist medical staff on deciding the best patient referral in order to reduce hospital stay for one-day surgeries [1, 2].

The rest of the paper is structured as follows. Section 2 presents the IoT infrastructure and Machine Learning background. Section 3 depicts the MAS Solution Architecture using IoT. Section 4 shows the application evaluation and results. Finally, Sect. 5 concludes the paper and present some future works.

2 Background

This section presents the fundamental concepts and technologies used in this article to understand the proposal. The topics internet of things, the XMPP communication protocol, machine learning and multi-agent systems will be addressed.

2.1 Internet of Things (IoT)

The Internet of Things is the term used to represent the digital interconnection among physical devices used in daily life throughout the world that until then could not interact via the Internet.

These IoT devices have been applied in areas such as industry, agriculture, health, computing and urban areas. Considering that the prices for the acquisition of these devices have become accessible to the final consumers, this technology is being incorporated into people daily life.

Among these devices, wearable devices, which are those that can be worn or mated with human skin to continuously and closely monitor an individual's activities, without interrupting or limiting the user's motions, have been in the academic community and industry attention [3], as shown in Fig. 1.

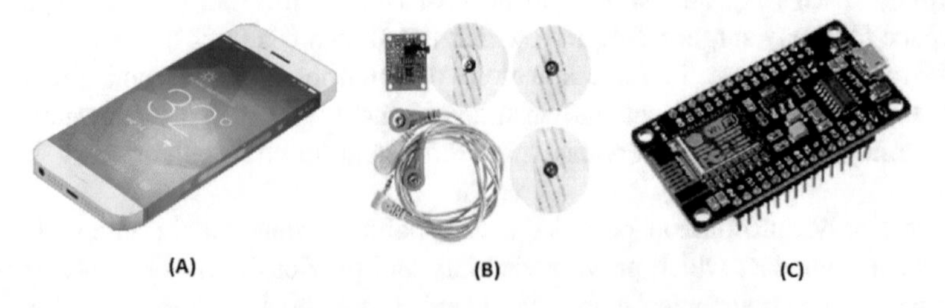

Fig. 1. (A) MQTT App, (B) AD8232 and (C) NodeMCU.

Device (A) is a smartphone that contains an APP which allows transferring collected data from the device to some server in the cloud. (B) represents body sensors that catch vital signals and when connected with (C), a kind of board with microprocessor, can send data (signals) to another device.

2.2 Extensible Messaging Presence Protocol (XMPP)

To interconnect the IoT devices in a network it was necessary to develop or improve lightweight communication protocols due to the scarcity of resources. Among these protocols came the XMPP protocol that was first named Jabber in 1999 and standardized by the Internet Engineering Task Force [14].

XMPP (Extensible Messaging and Presence Protocol) is a protocol based on Extensible Markup Language (XML) and intended for instant messaging (IM) and online presence detection. It functions between or among servers and facilitates near-real-time operation. The protocol may eventually allow Internet users to send instant messages to anyone else on the Internet, regardless of differences in operating systems and browsers. In addition, it has also been used in VoIP and IoT applications.

2.3 Machine Learning

One goal of Machine Learning (ML) - often also referred to as Data Mining or Predictive Analysis - is to teach machines (software) to carry out tasks by providing them a couple of examples (how to do or not do the task) [18]. Of course, it is not a brand-new field in itself. Its success in recent years can be attributed to the pragmatic way of using rock-solid techniques and insights from other successful fields like statistics.

There are several implementation of ML algorithms, and it's important to choose which one best fits on the specific problem, as shown in Fig. 2.

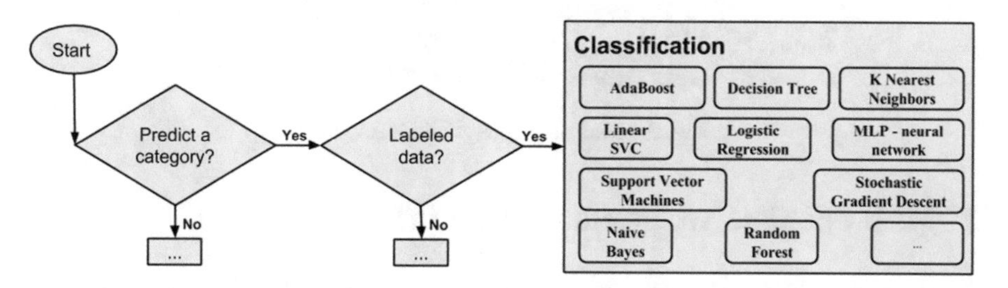

Fig. 2. Choosing the right estimator [9].

Figure 2 presents a fragment of the work flow which data scientists will be able to choose the best algorithm. The full "map" is shown in [9]. From the "START" point it's possible to reach the best model to apply. In this article, as we are using labeled data and intend to predict a category, we used classification estimators and compared the accuracy among them.

2.4 Multi-agent Systems (MAS) Applied to EHealth

In artificial intelligence research, agent-based systems technology has been hailed as a new paradigm for conceptualizing, designing, and implementing software systems. According to Russel [4], an agent is "anything that can be viewed as perceiving its environment through sensors and acting upon that environment through actuators.". A multi-agent system (MAS) is a network of software agents that is loosely coupled and interact to solve problems that are beyond the individual capacities or knowledge of each problem solver [5]. Considering this, Neto [11] proposed a Multi-Agent System using MQTT, IoT devices and Fuzzy Logic to infer the level of hypertension of a system user. Figure 3 shows the elements of this Multi-Agent System.

The Mobile Agent works catching the patient's vital signs. The Processing Agent can process information through fuzzy logic and infer how the patient's blood pressure is going, based on four parameters - systolic blood pressure, diastolic blood pressure, age and body mass index, and indicates some abnormality. The Monitoring Agent gets the process result which is shown on an output device. Publish/Subscribe Pattern allows the connectivity of this environment. In the present work we intend to implement new elements in the architecture proposed in [11]. The code of such implementation is available at Github[1].

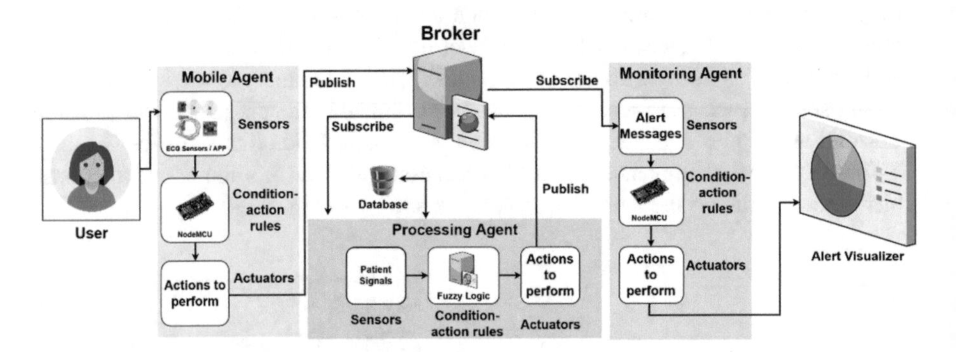

Fig. 3. Multi-Agent System(MAS) [11].

3 MAS Solution Architecture Using IoT

As stated by Russel [4], the task environment involves the PEAS (Performance, Environment, Actuators, Sensors) that should be described when designing an agent. The architecture proposed consists in three agents, as described on Table 1.

The architecture proposed by Neto [16] uses Fuzzy Logic as an artificial intelligence implementation. Besides, the connectivity uses Publish/Subscribe protocol to provide connectivity. The evolution of the embedded multi-agent architecture that we propose for our MAS/IoT solution is presented in Fig. 4.

[1] Available at: https://github.com/afonsoblneto/eHealth.

Table 1. PEAS description of the task environment

Agent type	Performance measure	Environment	Actuators	Sensors
Mobile	Data sent correctly	Hospital	XMPP client	Mobile APP
Processing	Refer post-operative patients accuracy	Cloud	Pub/Sub topic	XMPP
Monitoring	Availability time	Monitor centre	Mobile phone screen/Monitor screen	Pub/Sub topic

The elements of the current architecture implemented are introduced by [16], they are represented by rectangles with black solid lines, such as "RDBMS", "Mobile APP" etc. The elements with black dashed lines ("Smart Watch", "Smart Bands" and "NoSQL") aren't implemented yet. The two elements with red dashed lines ("Machine Learning" and "XMPP") are the ones which we propose to implement. We have chosen Machine Learning due to the variety of reliable algorithms, such as those implemented in Scikit-Learn library [9], which has several classification algorithms in order to infer the best option about where to send the post-operative patients. Besides, we used XMPP as a connectivity option in order to allow the Mobile Agent to send information about patients to the Monitoring Agent.

Is this work we propose an evolution of the previous architecture considering that the Mobile Agent that is encapsulated on a physical device (Smartphone) through an APP with which the user can interact. The Processing Agent can process information through machine learning classification estimators and infer where to refer post-operative patients (Intensive care unit, home or general hospital floor), based on the following information: patient's internal temperature (°C); patient's surface temperature (°C); Oxygen saturation (%); last blood pressure measurement; stability of patient's surface temperature; stability of patient's core temperature; and stability of patient's blood pressure.

A database stores the received and processed data for analytics purpose. The Monitoring Agent gets the process result which is shown on an output device. The Publish/Subscribe Pattern and XMPP "connects" this environment. The code of all services and applications are available on GitHub[2].

4 Application and Results

In order to implement the MAS/IoT solution, we developed a mobile APP using XMPP and Android OS [8]. The input fields are (i) patient's internal temperature (°C); (ii) Patient's surface temperature (°C); (iii) oxygen saturation (%); (iv) last measurement of blood pressure; (v) Stability of patient's surface temperature; (vi) Stability of patient's core temperature; (vii) Stability of patient's blood pressure. The mapping of the numerical data and categorical data is available at github.

[2] Available at: https://github.com/afonsoblneto/post_operative_ml.

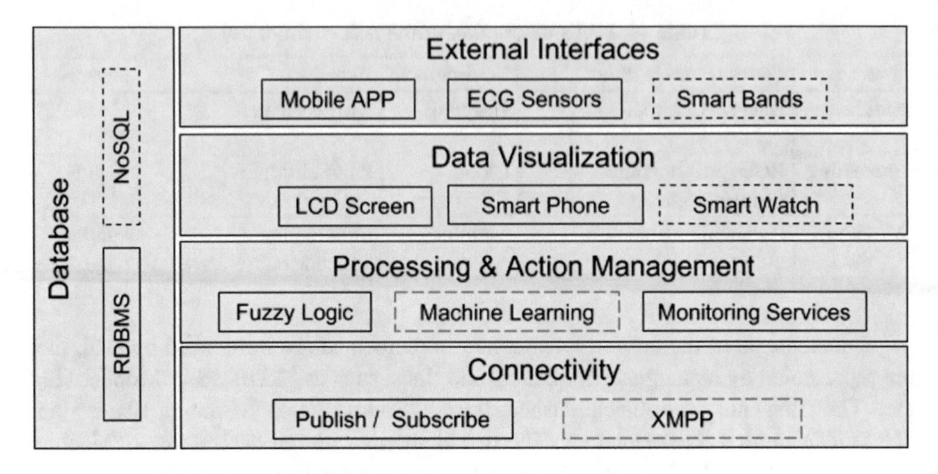

Fig. 4. Proposed architecture based in [16].

We used the following classification estimators, as show in Table 2.

Table 2. Classification estimators results

Estimator	Standard deviation	Accuracy % (average 72.57)	Runtime (in seconds)
AdaBoost	0.4	55.6	0.30
Decision Tree	0.9	76.8	0.02
K Nearest Neighbors	0.9	87.5	0.03
Linear SVC	1.0	70.2	0.52
Logistic Regression	1.0	70.2	0.03
MLP - neural network	0.3	88.8	4.52
Naive Bayes	0.6	38.9	0.01
Random Forest	0.8	78.9	1.16
Stochastic Gradient Descent	13.0	70.1	0.02
Support Vector Machines	0.5	88.7	0.11

The data set used in this work is the Post-Operative Patient Data Set, which is available at UCI Machine Learning Repository [10]. The original data set contains 90 rows. This is harmful for the estimating process due to overfitting. To get around this issue, we've added new rows, using Scikit Learn functions, in order to reach a total of 1890 rows, helping the estimators on Table 2 to improve their performance.

To check the Processing Agent performance, developed in Python [12], we used a Cross-validation approach (CV), which evaluates the estimator performance. A test set should still be held out for final evaluation, but the validation set is no longer needed when doing CV. In the basic approach, called k-fold CV, the training set is split into k smaller sets. The performance measure reported by k-fold cross-validation is then the average of the values computed in the loop. The scoring parameter used in CV was 'accuracy', using 95% confidence interval.

Table 2 also depicted that Stochastic Gradient Descent has the worst standard deviation - 13.0, while MLP - neural network, AdaBoost and Support Vector Machines have the best standard deviations - 0.3, 0.4 and 0.5, respectively.

As shown in Table 2, in the tests conducted with the estimators, the average accuracy was 72.57. The estimators MLP-NN, SVN, KNN, RF and DT had the accuracies greater than the average - 88.8, 88.7, 87.5, 78.9 and 76.8 respectively. The other estimators (LR, L-SVC, SGD, AdaBoost and NB) had accuracies lower than the average and should be discarded.

Considering the runtime over the best to worst estimator accuracy, as shown in Table 2, MLP-NN has the worst runtime, 4.52 s, followed by RF, 1.16 s. The remaining estimators had a runtime below 1 s.

Considering these two points, we recommend the SVN estimator to provide a high accuracy (88.7%) and low runtime (0.11 s), similar to the best accuracy from MLP-NN estimator with a lower runtime. The KNN estimator should be another good option, accuracy of 87.5% and runtime of 0.03 s. Then, considering the accuracy and runtime obtained, the proposed architecture should help a surgery center to improve the post-operative patient control.

Considering the input data, the SVM estimator referred 1280 patients to the hospital floor, 480 patients were referred home and 40 remained in the ICU.

5 Conclusion and Future Works

The use of technologies aiming health monitoring of hospitalized patients has grown in recent years, using novelties such as IoT devices in conjunction with machine learning techniques. This article shows an implementation of a multi-agent technology solution where, from the data collection of post-operative patients, points out the most appropriate form of referral for them.

The proposed implementation extends the one presented in [16], with the innovation of the application and evaluation of estimators based on machine learning techniques, instead of Fuzzy Logic classifiers. The best evaluated model, Support Vector Machines - SVM, reached an accuracy of 88.7% and a runtime of 0.11 s.

Considering that the dataset used initially is not large enough to train the classification algorithms, as future works we intend to use a bigger dataset to check the estimator's performance, in terms of runtime and accuracy.

References

1. Jenkins, K., Grady, D., Wong, J., Correa, R., Armanious, S., Chung, F.: Postoperative recovery: day surgery patients' preferences. Br. J. Anaesth. **86**, 272–274 (2001)
2. Haghi, M., Thurow, K., Stoll, R.: Wearable devices in Medical Internet of Things: scientific research and commercially available devices. US National Library of Medicine National Institutes of Health, January 2017

3. Pang, Z.: Technologies and architectures of the Internet-of-Things (IoT) for health and well-being. KTH Royal Institute of Technology, Kista Sweden, vol. xiv, 75 p. (2013)
4. Russell, S.J., Norvig, P.: Artificial Intelligence: A Modern Approach, 3rd edn. Prentice Hall (2010)
5. Shoham, Y., Leyton-Brown, K.: Multiagent Systems "Algorithmic, Game-Theoretic, and Logical Foundations", rev. 1.1 (2010)
6. De Oliveira, J.L.S., Da Silva, R.O.: A Internet das Coisas (IoT) com Enfoque na Saúde. Tecnologia Em Projeção **8**(1), 77 (2017). Http://Revista.Faculdadeprojecao.Edu.Br/Index.Php/Projecao4/Article/View/824/726. Accessed 22 Jan 2019
7. Pinheiro, A.L.S., Andrade, K.T.S., Silva, D.O., Zacharias, F.C.M., Gomide, M.F.S., Pinto, I.C.: Gestão da Saúde: O Uso dos Sistemas de Informação e o Compartilhamento de Conhecimento para a tomada de Decisão. Texto Contexto Enferm **25**(3), E3440015 (2016). Http://Www.Scielo.Br/Pdf/Tce/V25n3/Pt_0104-0707-tce-25-03-3440015.Pdf. Accessed 22 Jan 2019
8. Blikoon. https://www.blikoontech.com/tutorials/android-smack-xmpp-introductionbuilding-a-simple-client/. Accessed 11 Dec 2018
9. The Scikit-Learn. https://scikit-learn.org/. Accessed 21 Jan 2019
10. Dua, D., Karra Taniskidou, E.: UCI Machine Learning Repository. University of California, School of Information and Computer Science Irvine, CA (2017). http://archive.ics.uci.edu/ml
11. Neto, A.B.L., Andrade, J.P.B., Loureiro, T.C.J., de Campos, G.A.L., Fernandez, M.P.: A multi-agent system using fuzzy logic applied to eHealth. In: Novais, P., Jung, J.J., Villarrubia, G., Fernández-Caballero, A., Navarro, E., González, P., Carneiro, D., Pinto, A., Campbell, A.T., Durães, D. (eds.) Ambient Intelligence – Software and Applications – 9th International Symposium on Ambient Intelligence 2018. Advances in Intelligent Systems and Computing, vol 806. Springer, Cham (2018)
12. The Python. https://www.python.org/. Accessed 21 Jan 2019
13. Moraes, O., Carlos, L.: Framework de Comunicação seguro e confiável para Internet das coisas usando o protocolo XMPP 2016. 91 p. Dissertação (Mestrado em Engenharia de Eletricidade)-Universidade Federal do Maranhão, São Luis (2016). https://tedebc.ufma.br/jspui/bitstream/tede/1689/2/LuanCarlosOliveira.pdf. Accessed 08 Jan 2019
14. The XMPP. https://xmpp.org/. Accessed 22 Dec 2018
15. Yu, W.P., Chen, Y., Duan, G.M., Hu, H., Ma, H.S., Dai, Y.: Patients' perceptions of day surgery: a survey study in China surgery. Hong Kong Med. J. **20**, 134–138 (2014)
16. Neto, A.B.L., Andrade, J.P.B., Loureiro, T.C.J., de Campos, G.A.L., Fernandez, M.P.: Fuzzy logic applied to eHealth supported by a multi-agent system. In: Barreto G., Coelho R. (eds.) Fuzzy Information Processing. NAFIPS 2018. Communications in Computer and Information Science, vol 831. Springer, Cham (2018)
17. Bowyer, A.J., Royse, C.F.: Postoperative recovery and outcomes - what are we measuring and for whom? Anaesthesia **71**, 72–77 (2016)
18. Coelho, L.P., Richert, W.: Building Machine Learning Systems with Python, 2nd edn. Packt Publishing (2015). ISBN 978-1-78439-277-2

A Computing Framework to Check Real-Time Requirements in Ambient Intelligent Systems

Roua Jabla[1], Amani Braham[1], Félix Buendía[2](✉), and Maha Khemaja[3]

[1] Higher Institute of Computer Sciences and Communication Techniques, University of Sousse, Sousse, Tunisia
[2] Universitat Politècnica Valencia, Valencia, Spain
fbuendia@disca.upv.es
[3] University of Sousse, Sousse, Tunisia

Abstract. Ambient Intelligent (AmI) systems are related to environments that usually demand real-time characteristics in their context-aware applications. These systems can require a timing response according to specific needs dealing with context changes and user interactions. The current work presents a computing framework able to check and measure those times required by ubiquitous systems, running AmI processes on mobile or networked devices. The proposed framework focuses on checking real-time requirements of applications which can monitor and recognize human activities. This framework is then applied to two examples of well-known datasets related to Human Activity Recognition (HAR) and it supports the implementation of a mobile application which allows researchers to measure those latency times involved in HAR processes. The computation cost of training and prediction processes for these datasets is measured using such application and the obtained results highlight the critical effect of smartphone performance when processing sensor-based data and, specially, during the elaboration of online training models in this context.

Keywords: Computing framework · Real-time requirements · Context-aware applications · Ubiquitous mobile systems · Human Activity Recognition

1 Introduction

Nowadays, there are several areas that usually demand real-time characteristics in their context-aware applications. Ambient Intelligent (AmI) systems are part of these areas since they can require a timing response according to specific needs dealing with context changes and user interactions among others. The current work presents a computing framework able to check and measure those times required by ubiquitous systems, running AmI processes on mobile or networked devices.

AmI systems are built upon several aspects which require sensitive and responsive actions when users interact with them. These systems are commonly based on embedded and network devices that introduce communication issues which have to be managed. They also introduce context-aware factors which need to be recognized and processed, adapting their response to the user circumstances and environmental

P. Novais et al. (Eds.): ISAmI 2019, AISC 1006, pp. 19–26, 2020.
https://doi.org/10.1007/978-3-030-24097-4_3

features. All these aspects lead to the necessity of dealing with real-time requirements in AmI systems and checking their response time [1]. Moreover, the ubiquitous configuration of these systems introduces low performance devices and communication delays that along with the need to incorporate consuming-time reasoning and recognition algorithms, justify the proposal of computing frameworks which can measure AmI response times and check them in several contexts. Some frameworks have been designed to deal with the need to support real-time software in ubiquitous computing environments such as RCSM (Reconfigurable Context-Sensitive Middleware) [2], AmbientRT [3] an example of real time operating system for mobile wireless sensor networks, SmartUM [4] an intelligent ubiquitous middleware for smart cities, Marches [5] a context-aware reflective middleware for building real-time embedded systems, Campus [6] middleware for automated context-aware adaptation at runtime, or R2TCA [7] for developing reconfigurable real-time context-aware applications. These frameworks have been applied in multiple contexts such as SmartUM addressed to smart cities or R2TCA for baggage handling. The current work is focused on checking real-time requirements of applications which can monitor and recognize human activities. This context is also known as HAR (Human Activity Recognition) and there are several initiatives oriented towards sensor-based activity recognition of multiple user in smart home environments [8], the use of motion sensors for activity recognition in an ambient-intelligence scenario [9] or benchmark datasets for HAR and AAL (Ambient Assisted Living) [10]. In this work, a computing framework is proposed to examine some examples of human activities which can be performed within an AmI system and check their recognition under strong timing constraints.

The remainder of the work is structured as follows. The second section describes the framework architecture that supports the development of AmI related applications to check response times in a HAR context. Section 3 presents a case study addressed to test such computing framework in well-known examples of datasets. Finally, Sect. 4 draws some Conclusions and further works.

2 Computing Framework

2.1 HAR Processes

HAR has been a very active research area since late 90s [11] and the incorporation of wearable sensors has increased this interest [12]. Wearable sensor-based technology offers an opportunity to steadily control human activity through various sensors attached to the human body. With the ever increasing technological advances of smartphones, a new HAR trend has put forth an interesting idea, which tied to a particular use of smartphones as sensing devices due to the various on-board sensors in smartphones, ranging from accelerometer to magnetometer sensors. Shoaib et al. [13] looked into the role of three smartphone sensors (accelerometer, gyroscope and magnetometer) when solving the problem of inferring human activity. This work showed that the smartphone position, the performed classifier and the activity being recognized have an impact on the role as well as the performance of these sensors in the activity recognition process. Thereby, the ubiquity of embedded sensors on mobile phones has contributed and succeeded in facilitating the process of gathering sensor data related to a variety of physical activities performed by users [14]. Besides, HAR processes demand more than sensing the surrounding physical environment as well as the user behavior, and collecting the corresponding meaningful measurements. Indeed, these applications have to infer semantic outcomes in the sort of physical activities from gathered sensor measurements. To contribute significantly to this overall target, HAR processes have resorted to the use of Machine Learning (ML) algorithms, which strive for analyzing and classifying physical world information according to training patterns in a transparent way. There are multiple ML algorithms that are applied in HAR contexts [14]. Table 1 shows four algorithms, namely Naïve Bayes (NB), k-nearest neighbor (KNN), C4.5 and Random Forest (RF), which have been used in practical activity recognition problems on mobile devices.

Table 1. List of classification algorithms.

ML algorithms	Type	Accuracy	Main feature
Naïve Bayes (NB)	Bayesian	Low	Easiness in implementation
k-nearest neighbor (KNN)	Instance-based	Medium	Robustness against a noisy training data
C4.5 (DT)	Decision Tree	Medium	Treatment of numerical and categorical data
Random Forest (RF)	Classifier ensembles	High	Prevention from over-fitting

2.2 Framework Architecture

The proposed computing framework is based on a middleware architecture following up a client-server approach whose components run different HAR processes. On this account, the entire process – from activity sensing to activity inference - is done on mobile phones in real-time. Figure 1 shows a high-level view of this architecture which is structured in three basic modules. The *Activity Sensing* module is in charge of collecting raw data from built-in sensors and then storing them in the *Real test* dataset. In this case, body-worn sensors such as accelerometer or gyroscope are used to record user's movements and gestures. Once a set of sensor raw data has been collected, the system moves on to the next module that deals with the *Training model*. This training module behaves as a preparation step to model each physical activity, which is annotated in the training dataset with specific activity patterns. It can be performed either in the same mobile device (online training) or using trained models from a back-end server (offline training). The third module, called *Activity Inference* uses the activity patterns previously obtained in the training phase to predict and detect user states that can be labelled and stored in the server side for their further analysis. The prediction phase submits the "Real test dataset" into the training model for inferring semantic outcomes to the collected, yet unlabeled raw data samples on smartphone. Lastly, the labeled results are sent and backed up to the server in order to enable their further analysis and enrich new training datasets.

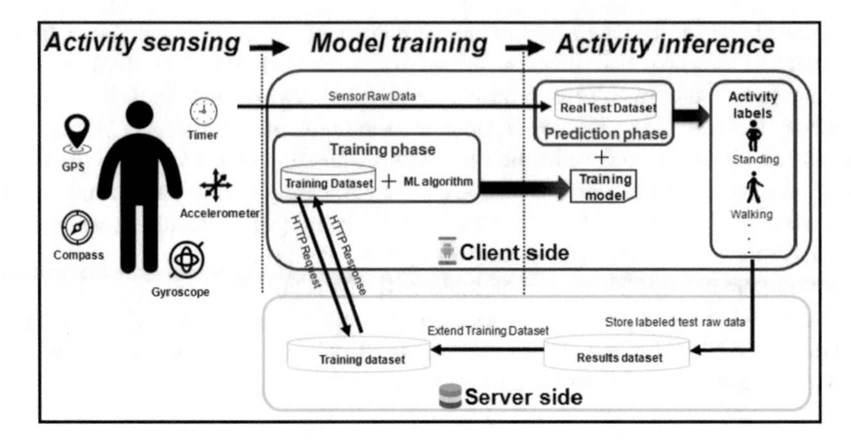

Fig. 1. Sensor-based HAR framework architecture.

2.3 Time Computation

The final goal of the proposed computing framework consists in obtaining the latency times associated to the HAR processes aforementioned. Figure 2 outlines the business process model that represents these processes and the set of tasks required to build our framework. They are organized into two main categories. The first category includes tasks that are executed during the sensing, training and prediction phases. The second one covers those tasks used for identifying computational costs in each activity recognition phase to make decisions. Sensors can trigger HAR processes when the acquisition begins. After that, a time measurement is launched through a "calculate latency time" task while a set of activities is instantiated. The treatment of this set of activities starts by calculating the time spent on collecting raw sensor values and then, the computational cost of training and prediction tasks is computed.

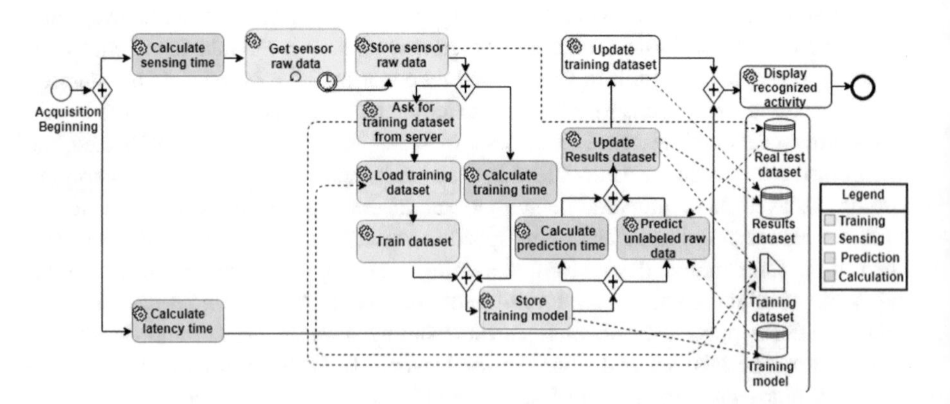

Fig. 2. Diagram illustrating HAR processes and their time computing.

3 Case Study

In this section, we present a case study that describes the application of the proposed framework in a HAR context. So, we first describe the datasets featuring such context. Then, a client-server implementation of the framework enables the analysis of smartphones' performance and its computational power when performing HAR processes by measuring their latency times for each phase. Eventually, our experimental results are presented and discussed.

3.1 HAR Datasets

In order to obtain experimental results consistent with our proposed HAR framework, we consider two well-known datasets published in the UCI machine learning repository. These datasets are based on smartphone-based inertial sensors to collect data and they have been used in several research works.

- The HHAR (Heterogeneity dataset for Human Activity Recognition) Dataset [15] involves motion data gathered using tri-axis accelerometer and tri-axis gyroscope, which are pre-installed in mobile phones and smart watches. This sensor data portrays six different human activities including, *Sitting*, *Standing*, *Walking*, *Biking*, *Stair up* and *Stair down*.
- The HAPT (Human Activity and Postural Transition) Dataset [16] contains motion data taken from inbuilt tri-axis accelerometer and tri-axis gyroscope on smartphones. This gathered data is symbolized in six basic human activities, i.e. *Sitting*, *Standing*, *Lying*, *Walking*, *Walking upstairs* and *Walking downstairs*. This dataset also includes postural transitions that occurred between static activities such as standing, sitting and lying activities. Postural transitions are represented in the form of *Stand-to-Sit*, *Sit-to-Stand*, *Sit-to-Lie*, *Lie-to-Sit* and *Stand-to-Lie* transitions.

3.2 Application Overview

For the purpose of our experiments, we developed a mobile application for implementing the proposed framework that included tasks such as training and recognizing human activities such as those displayed on Fig. 1. Our application also allows measuring latency times of these tasks and it was built to inspect them using the HHAR and HAPT datasets aforementioned. Two smartphone models were used to test such application in terms of computing the times spent in online training tasks upon provided datasets, processing sensor data within a specific window size or predicting the user states from these trained values. The WEKA library was used to process dataset values by generating training models that allowed the application performing prediction tasks in a later stage. This library contained a set of ML algorithms written in Java language and some of them were selected to classify the accelerometer data collected and backed up as an Attribute Relation File Format (ARFF) file. Eventually, the Java System class was used to fetch the latency times spent during the training and prediction tasks.

3.3 Experimental Results and Discussion

The proposed application was designed and implemented with the purpose of assessing the smartphones' performance over the different tasks entailed in the HAR processes mentioned in previous sections. This assessment procedure was outlined in Fig. 2 and is based on carrying out a temporal computational analysis of the tasks involved in such processes. So, we have focused on two main timing factors:

- *Training time* that measures the entire period of time spent from querying the training dataset, loading it and then building the associated model.
- *Prediction time* that measures the time employed by the obtained training model for predicting user activities from raw data gathered in real-time.

Figure 3 shows a chart that displays the latency times which are computed for those training and predicting tasks, using the HHAR dataset. Three recognition algorithms (DT, NB and KNN in the x-axis) have been tested with two different smartphones, while the y-axis represents their computational cost measured in seconds.

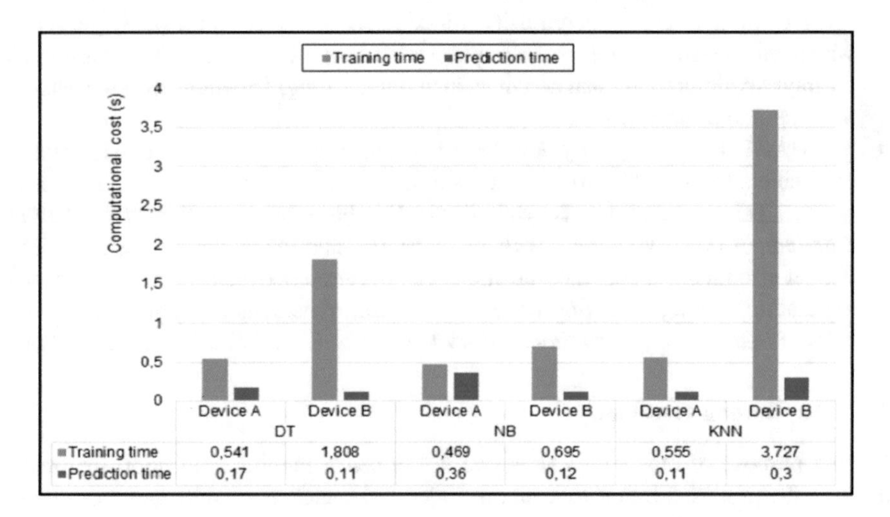

Fig. 3. Computational cost results for the HHAR dataset.

Similarly, Fig. 4 shows a chart with the computational cost associated to the test of the same recognition algorithms in the HAPT dataset. In general, latency times obtained during the training phases are higher in the HAPT case than the HHAR one, except the KNN algorithm for the device B. These higher values are due to the size and complexity of the HAPT dataset compared with the HHAR dataset. Computation cost displayed in both figures also indicates a better performance for device A compared with device B in training tasks. However, tests show almost similar prediction time results for HHAR and HAPT datasets using both smartphone types, due to the ground truth that online prediction stage requires much less device computational power than the training stage.

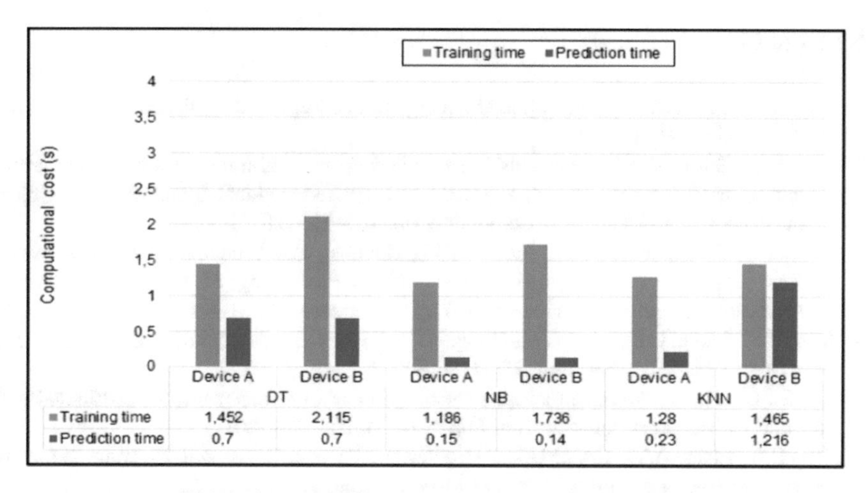

Fig. 4. Computational cost results for the HAPT dataset.

From the perspective of real-time HAR applications, the presented results point out to the fact that smartphone's performance level as well as dataset complexity has a critical impact on online training stage since the learning model is implemented on a mobile platform with limited resources. In this way, time spent on online training is rather directly proportional to the dataset complexity, the smartphone performance and those ML algorithms being used. Therefore, fulfilling timing requirements in this kind of applications demands to carefully shape the size of the sensor dataset to be trained directly on the mobile platforms or, alternatively, to use offline training techniques. Prediction times can be slightly influenced by the performance of the mobile platform where the activity recognition is applied so the selection of a suitable recognition algorithm is also important in this kind of applications.

4 Conclusions

The current work has presented a computing framework that allows researchers to check certain types of timing requirements in AmI systems and, specifically in those areas related to the human activity recognition (HAR). So, this framework has been implemented into a mobile application developing a set of experiments to compute the effect of smartphones on activity recognition processes. These experiments have been focused on assessing the computational cost of training and prediction tasks using different recognition algorithms and smartphone platforms. The main conclusions from the experiment results consist in taking into account the effect of the size and complexity of datasets used during the training stages along with the importance of selecting a suitable recognition algorithm that fits well with the performance of the mobile platform in which it can be executed. Further works will include the exploration of new techniques that assess the computational costs of training and prediction tasks in real HAR scenarios and the test of latency times associated to sensing tasks in such scenarios.

References

1. Augusto, J.C., McCullagh, P.: Ambient intelligence: concepts and applications. Comput. Sci. Inf. Syst. **4**(1), 1–27 (2007)
2. Yau, S.S., Karim, F.: Context-sensitive middleware for real-time software in ubiquitous computing environments. In: 4th Proceedings IEEE International Symposium on Object-Oriented Real-Time Distributed Computing, pp. 163–170 (2016)
3. Hofmeijer, T.J., Dulman, S.O., Jansen, P.G., Havinga, P.J.: AmbientRT-real time system software support for data centric sensor networks. In: Proceedings of the Intelligent Sensors, Sensor Networks and Information Processing Conference, pp. 61–66 (2004)
4. Jung, H.S., Jeong, C.S., Lee, Y.W., Hong, P.D.: An intelligent ubiquitous middleware for U-City: SmartUM. J. Inf. Sci. Eng. **25**(2) (2009)
5. Liu, S., Cheng, L.: A context-aware reflective middleware framework for distributed real-time and embedded systems. J. Syst. Softw. **84**(2), 205–218 (2011)
6. Wei, E.J., Chan, A.T.: CAMPUS: a middleware for automated context-aware adaptation decision making at run time. Pervasive Mob. Comput. **9**(1), 35–56 (2013)
7. Soumoud, F., Mohamed, R., Mohamed, K., Georg, F.: R2TCA: new tool for developing reconfigurable real-time context-aware framework-application to baggage handling systems. In: The 10th Proceedings of International Conference on Mobile Ubiquitous Computing, Systems, Services and Technologies, pp. 144–151 (2016)
8. Wang, L., Gu, T., Tao, X., Lu, J.: Sensor-based human activity recognition in a multi-user scenario. In: Proceedings of the European Conference on Ambient Intelligence, pp. 78–87 (2009)
9. Cottone, P., Re, G.L., Maida, G., Morana, M.: Motion sensors for activity recognition in an ambient-intelligence scenario. In: Proceedings of the IEEE International Conference on Pervasive Computing and Communications, pp. 646–651 (2013)
10. Amato, G., Bacciu, D., Chessa, S., Dragone, M., Gallicchio, C., Gennaro, C., Vairo, C.: A benchmark dataset for human activity recognition and ambient assisted living. In: The 7th Proceedings of the Ambient Intelligence-Software and Applications–International Symposium on Ambient Intelligence, pp. 1–9 (2016)
11. Foerster, F., Smeja, M., Fahrenberg, J.: Detection of posture and motion by accelerometry: a validation study in ambulatory monitoring. Comput. Hum. Behav. **15**(5), 571–583 (1999)
12. Lara, O.D., Labrador, M.A.: A survey on human activity recognition using wearable sensors. IEEE Commun. Surv. Tutorials **15**(3), 1192–1209 (2013)
13. Shoaib, M., Scholten, H., Havinga, P.J.: Towards physical activity recognition using smartphone sensors. In: Proceedings of the 10th Conference on Ubiquitous Intelligence and Computing, pp. 80–87 (2013)
14. Lara, O.D., Labrador, M.A.: A mobile platform for real-time human activity recognition. In: Proceedings of Consumer Communications and Networking Conference, pp. 667–671 (2012)
15. Stisen, A., Blunck, H., Bhattacharya, S., Prentow, T.S., Kjærgaard, M.B., Dey, A., Jensen, M.M.: Smart devices are different: assessing and mitigating mobile sensing heterogeneities for activity recognition. In: Proceedings of the 13th ACM Conference on Embedded Networked Sensor Systems, pp. 127–140 (2015)
16. Anguita, D., Ghio, A., Oneto, L., Parra, X., Reyes-Ortiz, J.L.: A public domain dataset for human activity recognition using smartphone. In: European Symposium on Artificial Neural Networks, Computational Intelligence and Machine Learning ESANN (2013)

A Ubiquitous Computing Platform for Virtualizing Collective Human Eyesight and Hearing Capabilities

Risa Kimura(✉) and Tatsuo Nakajima

Waseda University, Tokyo, Japan
{r.kimura, tatsuo}@dcl.cs.waseda.ac.jp

Abstract. Ubiquitous computing technologies have dramatically changed our daily lifestyles. A mobile phone that contains powerful computational and sensing capabilities allows us to access information anytime, anywhere. In the near future, we will wear mobile phones as glasses and clothing, and the wearable devices will be used to enhance our bodies' capabilities. In particular, the camera and microphone contained in such a device can be used as alternative eyes and ears for developing novel types of services. This study investigates the feasibility of sharing other people's body parts, particularly their eyes and ears, to build novel services. A platform that would enable a user to adopt anyone else's seeing and hearing capabilities would offer novel sharing economy services that would go beyond the traditional services that share people's belongings and would instead share the human body's capabilities. We have developed a prototype implementation of the platform, and we present the results of preliminary user studies to investigate the feasibility of our approach.

Keywords: Virtual reality · Sharing economy · Gaze-based gesture

1 Introduction

Ubiquitous computing technologies have dramatically changed our daily lifestyle. A mobile phone containing powerful computational capabilities allows us to access information anytime, anywhere. Since the phone also has various sensing capabilities, it can retrieve various information from the real world. In the near future, we will wear mobile phones as glass and clothing, and such a device will be used to enhance our bodies' capabilities [4]. For example, a Google Glass contains a mobile phone's capabilities but is designed in the shape of a pair of eyeglasses. This device reveals the possibility of enhancing our eyes and ears' capabilities[1]. In particular, the camera and microphone contained in the device can be used as other people's alternative eyes and ears.

These ubiquitous computing technologies enable us to build new types of services. For example, most people can connect to each other via mobile phones. People may act as sensors to retrieve various pieces of information about the real world [8].

[1] https://www.youtube.com/watch?v=4EvNxWhskf8.

P. Novais et al. (Eds.): ISAmI 2019, AISC 1006, pp. 27–35, 2020.
https://doi.org/10.1007/978-3-030-24097-4_4

Additionally, playing a fictional role in the real world may make it easier to motivate people to behave more sustainably [4]. In the near future, technology may change our sense of ownership over the human body [7], so a new type of sharing economy that shares parts of the human body will become feasible.

In this study, we explore the possibilities of such a new ubiquitous computing technology, and we present a distributed platform to virtualize collective human eyesight and hearing capabilities; the platform makes it possible to develop novel services that would allow users to adopt someone else's seeing and hearing capabilities through their wearable devices that contain camera and microphone functionalities. The platform adopts a user's gaze-based gestures to choose which eye view he/she would like to see and to direct the orientation of the head of the person whose vision the user has adopted. The study includes some preliminary user studies to discuss the feasibility of the prototype platform. The user studies allow us to extract interesting insights to understand the potential benefits and pitfalls of sharing collective human eyesight and hearing capabilities.

2 Related Work

A sharing economy, referring to the peer-to-peer-based sharing of access to goods and services, has recently attracted a great deal of attention [1]. The sharing economy covers a sprawling range of digital platforms and offline activities such as Airbnb, a peer-to-peer lodging service, and Uber, a peer-to-peer transportation network. The sharing economy typically uses information technology to provide individuals, corporations, nonprofits and governments with information that enables the optimization of resources by the redistribution, sharing and reuse of excess available goods and services.

"Jack-in" is a concept used for augmenting human capability and human existence [5]. The concept enables an immersive connection between humans and other artifacts or between humans. Our approach considered a distributed platform that uses collective human visual and auditory capabilities and that is based on this concept.

KinecDrone enhances our somatic sensation of flying in the sky [3]. A video stream captured by a drone is transmitted to a user's head-mounted display. While a user behaves as though he/she is flying in the sky while still in a room, he/she can watch the scene captured by the flying drone. Thus, the user feels like he/she is truly flying in the sky.

Crowdsourcing divides a task into various micro-tasks and asks multiple people to complete them. As shown in [8], humans can be used to monitor real-world information by acting as sensors, an approach that is based on crowdsourcing methods.

The proposed platform is novel because there are no previous approaches that have allowed us use collective human visual and auditory capabilities to build advanced ubiquitous computing services.

3 Design and Implementation

Figure 1 shows a typical use case scenario in which a user uses the proposed distributed platform. A user first specifies a region where there are people watching something that the user has an interest in (a). The user watches several people's viewpoints simultaneously in a virtual space and selects one of them to adopt his/her eye view (b). Then, the user selects and adopts one of these eye views that may contain something that he/she is interested in (c). The user turns his/her head to better see what he/she is interested in (d). A moving object appears to move the orientation of the selected person's head (e).

In this approach, having multiple eye views options is essential for a user to find things he/she is interested in (Fig. 1b). The user needs to be able to find a person who is close to what he/she is interested in (Fig. 1c). He/she can then direct the orientation of the selected person's head to make him/her better able to see things of interest to the user (Fig. 1e).

3.1 Design Approach

The proposed platform consists of the following four functionalities. We assume that each person is equipped with a wearable device containing a camera and microphone, typically wearable glasses. The current version of the platform uses a head-mounted display (HMD) and includes a camera and microphone on the front of the HMD (Fig. 2a). The HMD projects views captured by the cameras. However, the future version of the platform will use smart glasses to improve its feasibility for daily use.

Eye Gaze-based Gesture: One of the characteristics of our approach is to use gaze-based gestures for all controls in the platform. We defined the following five basic commands. The first command is employed by the user to select a target person by moving his/her eyesight from top to bottom (select command). The second command is used to return to the previous view by the user moving his/her eyesight from bottom to top (deselect command). The third command is employed by the user to choose another person whose eye view the user wants to adopt by the user watching the selected person and moving his/her the eyesight from top to bottom around the person (adopt command). The fourth command is used to change the current view to the randomly selected view of another person near the user by the user moving his/her eyesight from top to bottom in the current view (change command). The last command is used for removing a view that the user wants to replace by the user moving his/her eyesight to from bottom to top on the current view (replace command).

Finding an Appropriate Eye View: First, a user chooses a region on a map, as shown in Fig. 2b-1. Then, multiple eye views of people who are in the region are presented to a user in a virtual space. The user chooses one of views with the select command. The view of a selected person is shown instead of the user's current eye view to make him/her feel as though he/she has adopted the selected person's eye view. The user can then direct the selected person's view with his/her eye gaze, as shown below. If the user finds that the selected person is not an appropriate person, he/she chooses another person with the adopt command (Fig. 2b-2).

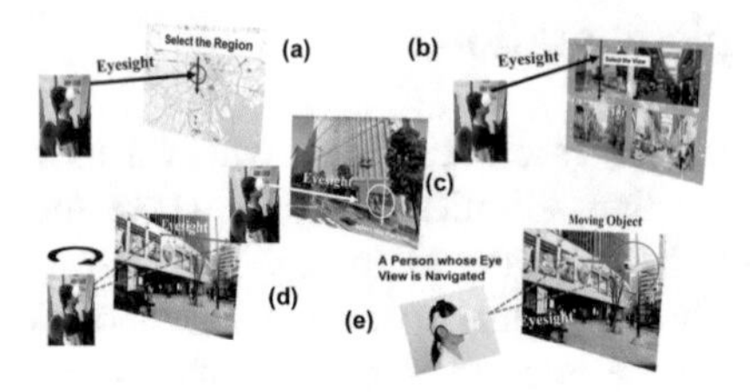

Fig. 1. A use case scenario

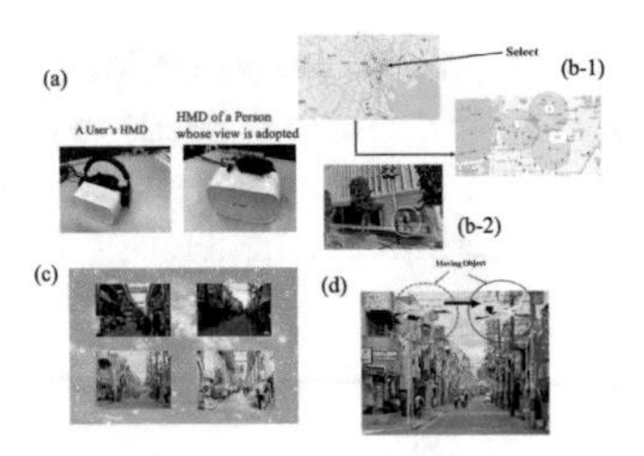

Fig. 2. An overview of the proposed platform: (a) A head-mounted display; (b) Finding an appropriate view (1) from a map, (2) by adopting another person's eye view; (c) Presenting multiple eye views; (d) Directing a selected person's eye view

Watching Multiple Eye Views: When showing multiple eye views, the views are shown in a virtual space, as shown in Fig. 2c. When there are too many views to be shown in the virtual space, four views are automatically selected. If one of views does not interest the user, he/she uses the gaze-based gesture for the remove command to indicate that the view is not necessary. Then, another view is shown instead of the removed view. Additionally, the change command can replace all views at the same time.

Directing a Selected Person's Head Orientation through the User's Eye Gaze: When a user wants to see an object at the edge of the current view, he/she must turn the selected person's head orientation toward the object. Our approach involves directing the selected person's head orientation by showing a moving/blinking object or/and making ambient sound in the 3D sound space, as shown in Fig. 2d.[2] When a user moves his/her eye gaze, a moving or blinking object appears or a sound is made in the direction the user wishes to move the orientation of the selected person's head.

3.2 Prototype Implementation

Figure 3 shows the software structure of the proposed platform that we have developed. The current platform can use an HMD or a public display to watch another

[2] People tend to notice a moving or blinking object when it appears in their peripheral vision [10].

person's view and uses Fove 0[3] or the Tobii Eye Tracker 4C[4] for tracking people's eye gazes. We currently attach a standard web camera and microphone to the HMD or a public display. The current prototype consists of the following four modules.

View Generation: This module generates a view that a user watches by using Unity. The view switches between the following three modes. The first mode shows a map, the second mode presents multiple eye views, and the third mode shows a selected eye view. The view switches modes through gaze-based gestures.

Head Direction: This module superimposes a moving or blinking object on a scene captured through a camera by using Unity. It can also use 3D sound to direct a person's head orientation.

Gaze-Based Gesture Management: This module extracts a person's eye gaze through an eye-tracking device. The movement of the eye is translated into the respective gesture commands.

Spatial Database: This module is used to find people who are in a specified region.

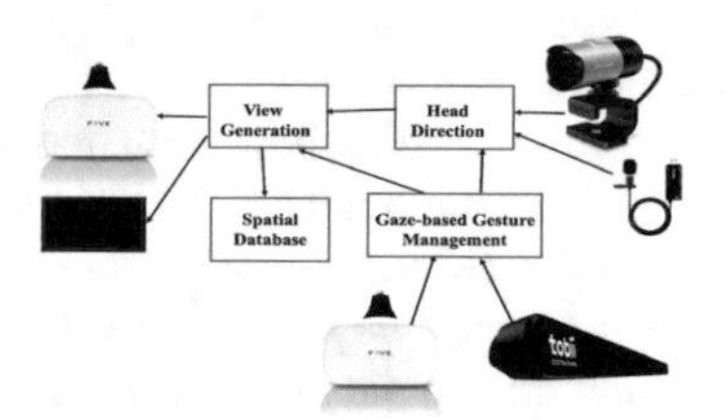

Fig. 3. An overview of a prototype implementation

4 A Preliminary User Study

In the user study, we examined the feasibility of the proposed approach using two experiments. In the first experiment, we asked participants to use our platform and conducted semi-structured interviews with the participants to extract some insights into the current approach. The second experiment investigated the effect of gaze-based head direction in directing the orientation of a selected person's head.

4.1 Prototype Implementation

At the beginning of this experiment, we explained the background and motivation of our research. Then, we presented some scenarios to help participants understand how to use the platform, and we asked them to perform the scenarios with the prototype platform. We hired 16 participants (age m = 24.8, sd = 8.4, 14 males) for the experiment.

[3] https://www.getfove.com/.

[4] https://www.tobii.com/.

In the interviews after performing the scenarios, we mainly asked about the following three topics. The first topic was the gaze-based gestures and the gaze-based direction of a selected person's head orientation. The second topic was the effect of adopting other persons' eye view. The final topic was whether participants would consider using the platform in their daily lives.

All participants said the gaze-based gesture was natural and easy to use, and it was easy to remember how to use respective gesture commands. Additionally, they found gaze-based head direction to be useful for controlling the orientation of a selected person's head just by moving his/her eyesight. However, one participant said, "*For performing gestures, moving my eyesight should be minimal.*" Another participant claimed that "*When using the platform for a long time, I felt my eyes get tired.*"

All participants answered that adopting other persons' eye views was enjoyable and useful because they could easily investigate the environment from multiple angles, but one participant claimed "*Adopting an eye view is dangerous in terms of privacy because I can easily know whose view is now adopted. Selecting a person's view randomly may be more preferable.*" Another participant said, "*Adopting multiple persons' views consecutively offers serendipitous thinking to expand my creativity.*" Additionally, one participant said, "*The eye views of foreign people who visit Tokyo might offer me serendipitous ideas about places that are well-known to me.*"

One important concern of the approach is how participants felt about privacy issues. Their comments suggested that the issue is not serious if the anonymity of a person who offers his/her eye view is preserved. However, one participant claimed that "*I think that it is necessary to consider the privacy of the person in the field of vision; I don't want to show the information presented on my mobile phone.*" Another participant said, "*I wouldn't like someone controlling my vision while I am reading books, talking to people, or relaxing at home.*" Additionally, in terms of using the platform in daily life, one participant stated that "*I may confuse whether I'm now watching my view or others' views. So, the platform should clearly indicate whether the current view is not my view.*"

4.2 Direct the Orientation of a Selected Person's Head

The second experiment aimed to understand the effects of the gaze-based head direction used in our approach; we hired 18 participants (age m = 24.2, sd = 8.1, 2 females) for this experiment. In this experiment, we presented a moving or blinking visual object at the edge of a participant's vision so that the prototype platform would direct his/her head toward that direction. We also used 3D sound to direct a participant's head orientation. The experiment was configured in the following five patterns: moving visual object (*MO*), blinking visual object (*BO*), 3D ambient sound (*S*), *MO* + *S*, and *BO* + *S*. In the experiment, we considered the following three situations as shown in Fig. 4 when directing the orientation of a person's head: the first situation is one in which the participant is walking and looking around (S1), the second is one in which the participant is looking his/her Timeline on Facebook (S2), and the third is one in which the participant is waiting for the signal to change at a crossing (S3). After the experiment, we surveyed the participants' responses to the five patterns using a five-point Likert scale (5-Induced, 4-Induced to some extent, 3-Cannot say either way, 2-Induced sometimes, 1-Not induced) for each situation. We also conducted an interview with each participant to ask the reasons for low scores.

When a person is walking and looking around, his/her eye view focuses on the direction that he/she is facing (S1). The results of the average scores m(sd)[5] in the survey are *MO* 3.2(1.1), *BO* 3.4(1.0), *S 3.4*(1.3), *MO + S 4.7*(0.5) *BO + S 4.4*(0.8). In discussing the lower scores in the cases of *MO* and *BO*, the participants explained that the moving or blinking objects were lost against the background. In particular, some participants said that the blinking object was hard to distinguish from the dappled sunlight in the scene, and the moving object was hard to distinguish from moving leaves. Additionally, using the 3D ambient sound showed diverse effects on the head direction because the accuracy of the sound direction was perceived differently by different participants. However, the visual object combined with a sound was generally good for directing the head orientation because the sound came to signify the appearance of the object.

When a user was watching his/her Timeline on Facebook (S2), he/she usually focused on looking at the Timeline. The results of the average scores m(sd) are *MO* 3.0 (1.1), *BO* 3.2(1.0), *S* 3.2(1.3), *MO + S* 4.4(1.0), *BO + S* 4.2(1.2). A moving/blinking object combined with the sound also seemed preferable, as it did in the first scenario. However, some participants told us they usually look at their Facebook timeline while listening music with headphones. Therefore, any use of sound as a method for notification should be carefully designed.

When a person is waiting for a signal to change at a crossing (S3), he/she may be relaxed and may not focus on looking at anything. The results of average scores for this scenario m(sd) are *MO* 2.5(1.0), *BO* 3.2(1.0), *S* 3.3(1.4), *MO + S* 4.3(0.9), *BO + S* 4.6 (0.7). In this case, the visual notification usually did not work well because some participants claimed that several cars were moving in front of them, so the moving cars overrode their awareness of the appearance of the visual objects. The MO(+S) score is lower than the BO(+S) score because the moving objects are more distinguishable from moving cars.

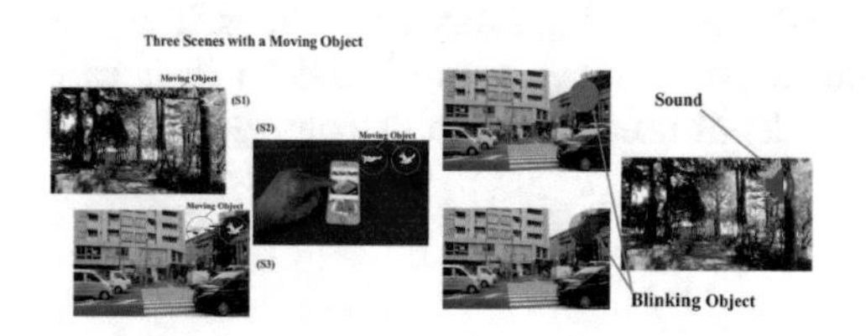

Fig. 4. Three scenes used in the experiment

As shown above, using sound combined with visual objects usually had a better effect. However, in daily urban live, people tend to listen music with headphones. Directing the orientation of people's heads based on sound should be carefully designed so as to not to disturb their music and thereby make the experiment uncomfortable.

[5] m: mean, sd: standard deviation

5 Possibilities and Challenges

Our prototype system allows us to choose a view anonymously to preserve people's privacy, but we still need to consider the following two issues. The first issue is the possibility of identifying a person from his/her surroundings as seen in the views. For example, if the person looks at his/her belongings, someone may see who he/she is. The second issue is that the view may be used to know someone's current situation. For example, if a view shows a person meets another person in secret, that fact may be disclosed to others.

We have considered only human eye views in the current approach. As shown in [9], a robot's view offers opportunities to sightsee in a remote place, to attend a conference, or to meet with faraway family members. In a scientific fiction story named Beatless [2], the number of humanoid robots becomes almost equal to the number of humans, and the robots are used to replace humans in performing various boring tasks. In this case, these robots may offer multiple eye views because controlling a robot's view is easier than controlling a human's view.

The proposed approach requires diving into a VR space for accessing multiple views. In [6], we showed that presenting multiple eye views offers a new possibility for reflective thinking. However, for using this approach in our daily lives, a more lightweight method to access the VR space is desirable. One approach is to present multiple views in the real space by using augmented reality technologies, but the views should be carefully shown in the real world without increasing a user's cognitive load and disturbing his/her currently tasks.

6 Conclusion

This paper proposed a distribution platform for developing sharing economy services that aggregate humans' visual and auditory capabilities. The platform enables users to guide through multiple people's eye views based on their gaze-based gestures. The paper showed the results of user studies to demonstrate the platform's feasibility.

References

1. Hamari, J., Sjöklint, M., Ukkonen, A.: The sharing economy: why people participate in collaborative consumption. J. Assoc. Inf. Sci. Technol. **67**(9), 2047–2059 (2016)
2. Hase, A.: Beatless. Kadokawa Publishing, Tokyo (2012)
3. Ikeuchi, K., Otsuka, T., Yoshii, A., Sakamoto, M., Nakajima, T.: KinecDrone: enhancing somatic sensation to fly in the sky with Kinect and AR.Drone. In: Proceedings of the 5th Augmented Human International Conference (2014). Article 53
4. Ishizawa, F., Sakamoto, M., Nakajima, T.: Extracting intermediate-level design knowledge for speculating digital–physical hybrid alternate reality experiences. Multimedia Tools Appl. **77**(16), 21329–21370 (2018)
5. Kasahara, S., Rekimoto, J.: JackIn: integrating first-person view with out-of-body vision generation for human-human augmentation. In: Proceedings of the 5th Augmented Human International Conference (2014). Article No. 46

6. Kimura, R., Nakajima, T.: Sharing collective human's eyesights towards reflective thinking. In: Proceedings of the 17th International Conference on Mobile and Ubiquitous Multimedia, pp. 341–349 (2018)
7. Kilteni, C., Maselli, A., Kording, K.P., Slater, M.: Over my fake body: body ownership illusions for studying the multisensory basis of own-body perception. Front. Hum. Neurosci. **9**, 141 (2015)
8. Liu, Y., Alexandrova, T., Nakajima, T.: Using stranger as sensors: temporal and geo sensitive question answering via social media. In: Proceedings of the 22nd International Conference on World Wide Web, pp. 803–814 (2013)
9. Tsui, K.M., Desai, M., Yanco, H.A., Uhlik, C.: Exploring use cases for telepresence robots. In: Proceedings of the 6th International Conference on Human-Robot Interaction (HRI 2011), pp. 11–18 (2011)
10. Weinschenk, S.: 100 Things Every Designer Needs to Know About People. New Riders Press (2011)

A Recurrent Neural Network Approach to Improve the Air Quality Index Prediction

Fabio Cassano(✉), Antonio Casale, Paola Regina, Luana Spadafina, and Petar Sekulic

Omnitech Research Group, Bari, Italy
{fabio.cassano,antonio.casale,paola.regina, luana.spadafina,petar.sekulic}@omnired.eu

Abstract. Every year, cities all over the world face the problem of the air pollution. In particular seasons, such as the winter, the levels of bad particles coming from the industrial and domestic heating systems increase the risk of pulmonary diseases. Thus, for both the city majors and citizens, it is important to understand and predict the air pollution levels in advance to safe guard the health. Modern forecasting systems are able to alert the population in advance only about the general weather condition, while the air quality information are almost not considered at all. The reasons are manifold and they mostly depend by the difficult that the modern systems have to generalize the problem and correct elaborate the data coming from the sensors. In this paper we address the problem of forecasting the bands of the different air pollutants according to the Air Quality Index in the Apulia region. Using two different Recurrent Neural Network models, we have performed two tests to prove that it is possible to predict the level of the pollutants in a specific area by using the data coming from the surrounding area. By using this approach on both the weather and air stations on the territory it is possible to have alerts many days ahead on the pollution levels.

Keywords: Air Quality Index · Recurrent Neural Network · Blind prediction

1 Introduction

The air quality level monitoring is receiving an increasing attention by researchers all over the world. Breathing clean air is essential for the human health and well-being, however, many people around the world live in places where the constant air pollution expose them to an higher risk of pulmonary disease or lung cancer [1]. Among the different pollution particles present in the air, the small one such as the $PM_{2.5}$ and the PM_{10} are considered among the most dangerous for the human health causing pulmonary cancer. In big cities close to industrial manufacture poles, the levels of those elements are so high that people moving around need to wear protective masks to safeguard their health.

P. Novais et al. (Eds.): ISAmI 2019, AISC 1006, pp. 36–44, 2020.
https://doi.org/10.1007/978-3-030-24097-4_5

There are many alert systems that allow the population to know in advance the levels of small pollutants particles, however, those are capable to predict with few advance when those level are going to exceed the limit. The overall presence of harmful particles in the air depends not only by the industrial waste, the vehicular traffic, or the urban heating systems but also by the weather conditions (such as the wind, the humidity of the air, the rain etc.). As a matter of fact, while the polluted air moves according to the direction of the wind from a city to the nearby ones expanding the discomfort and the population health-related issues, the rain “cleans” the air lowering the bad particles level. Having a good prediction of the quality of the air has manifold benefits, such as:

1. Allowing the population to be warned in time of the increasing number of pollution particles;
2. Allowing the local governments to adopt social strategies (such as limiting the traffic to a specific type of vehicles) and,
3. Allowing companies to schedule their production to avoid the excess of pollution in the air.

In this paper we address the problem of forecasting the air quality by analyzing the weather and the air pollution level of the Apulia (Italy) region. We developed two different models that, using modern Deep Learning techniques, forecast the level and the air quality (evaluated according to the Common Air Quality Index, or CAQI) considering the current weather conditions and the historical pollution data of a given place. Recurrent Neural Networks (RNNs), which are the state of the art method in sequence labelling problems, are used to predict the air quality level. We have performed two test: the first takes into account the data from 2012 to 2017 and randomly extracts the days for training and test. The second, using the same data, randomly chooses, one air station and considers it as test, while the others are used for training. In this way we prove that the algorithm is able to predict the air levels, without using the data coming from the chosen air station.

This paper is organized as follows. Section 2 describes the related works of the techniques applied to the air quality and the forecasting of the pollution. Section 3 describes the details of the proposed solution. Section 4 shows the results obtained by our tests, while Sect. 5 draws the conclusions and the future works.

2 Related Works

Nowadays Artificial Intelligence (AI) techniques are widely used in many field, from the medicine to the weather forecasting to the art. Thanks to the recent discoveries and the increasing computational power of modern personal computers, complex dataset can be elaborated in few time reaching an incredible accuracy. This can be also improved by mixing multiple approaches such as the evolutionary one and the machine learning [2,3]. Modern Machine Learning (ML) techniques such as the Artificial Neural Network, the Genetic Programming,

Deep Neural Network, etc. allow scientists from all over the world to forecast the level of harmful particles, such as $PM_{2.5}$ or the PM_{10} [4]. Those models also provide useful information on how the pollution moves from one area to another of a country. However, the forecasting accuracy rapidly decreases according the days ahead and the data used. Among the different neural network models, those with the Gated Recurrent Unit (GRU) and the Long Short Term Memory (LSTM) are used to predict values using time series [5]. However, only the usage of LSTM neural networks is currently being explored to predict the pollution in many cities [6].

To reach a good prediction level, all the ML techniques need valuable data to work with. There are many ways to retrieve environmental data: from the weather stations installed in different places of a given area, to the satellite image analysis or even by crowd-sourcing [7]. Most of the array sensor used to analyse the air are commonly installed both in the city center to track the level of bad particles from the transports, and close the industrial poles. The acquisitions are commonly scheduled in a fixed time span, however, many scientists have adopted innovative ways to get real-time data by moving through the area monitoring the level of pollutants [8]. Lastly, depending on the data acquisition rate, the ML model can be tuned to improve the prediction accuracy [9].

To easily define the overall quality of air pollutants, each nation of the world adopts a classification method called "Air Quality Index" or AQI. Its aim is to classify the air quality into different levels taking into account several parameters such as the density of micro particles like the CO_2, the O_3 etc. In general it takes into account the range of the different pollutants and classifies the level of the air into different categories. To each of those categories it is associated a "health implications" for the population [10]. This method allows a fast evaluation of the overall quality of the air, however, due to the different government laws, AQI ranges are not standardized among countries (and continents) all over the world [11]. As a matter of fact, both the level of pollutants and the number of the AQI groups depend on the nation and the standard adopted. Thus, it is possible that a "Good" AQI for a given country is considered "Moderate" for another one. This creates confusion and does not allow forecasting models developed in different countries to be easily adopted worldwide. Trying to predict the AQI index according to the different parameters is still an open question that drives researchers to use techniques such as the fuzzy logic to improve the forecast [12]. Having good results also improve people's lifestyle allowing them to spend less time at home [13].

3 The Proposed Solution

In this paper we present two tests performed on different RNN models. Both aim to the prediction of the CAQI level using data coming from weather and air stations. By using those models on each station, it is possible to predict the quality of the air and warn the population in advance for possible air pollution-related problems. In this way, each station become "smart" thanks to the integration of the those AI algorithms and the "freshness" of the acquired data.

RNN have proved to be very flexible and accurate in multiple contexts involving time series thanks to their ability to elaborate datasets. Among the different types, those that perform better use the Long Short Term Memory (LSTM) and Gated Recurrent Unit (GRU) neurons. Discussions on the differences between these two types of neurons have been explored in [14], where it has shown that the LSTM architecture performs better on some very specific ML problems (such as numeric dataset), while the GRU has a general better results on others (such as audio and some numeric one).

Table 1. Statistics of measured values. Unit, range, maximum, minimum, mean, and standard deviation values.

Element	Unit	Range	Mean	Standard deviation
NO_2	$\mu g/m^3$	$[0.0, 387.0]$	22.8	13.2
PM_{10}	$\mu g/m^3$	$[0.0, 278.0]$	35.9	23.9
O_3	$\mu g/m^3$	$[2.0, 199.0]$	92.2	25.1
Tmin	°C	$[-1.3, 41.2]$	17.9	6.9
Tmax	°C	$[-1.2, 42.3]$	18.6	7.1
Humidity	%	$[12.6, 100.0]$	66.6	17.2
Wind speed	m/s	$[0.0, 9370.0]$	2.3	37.3
Wind direction	°	$[0.0, 359.0]$	209.8	101.9
Atmospheric pressure	hPa	$[964.1, 1027.9]$	1004.8	6.9
Day of year	*int*	$[-1, 1]$	-	-
Day of week	*int*	$[1, 7]$	-	-

In order to get a good prediction accuracy, we have used two different datasets at the same time: the former related to the weather conditions, the latter related to the quality of the air in the different places of the Apulia region. Both of them provide near-real-time data, freely available on the ARPA Apulia web site[1,2]. The first dataset contains information about the geographical position of the weather stations and what kind of sensors are available. In addition, each entry contains the timestamp of the acquisition and the values coming from the sensors (such as the wind direction, the humidity etc.). The second dataset is about the general condition of the air quality. Each entry contains information about the major air pollutants including the $PM_{2.5}$, PM_{10}, CO_2, NO_2 as well as the timestamp of the acquisition. Statistical information about the dataset values and how those have been normalized are reported in Table 1. The first step to get the RNN correctly trained has been to clean the datasets by reducing the number of features the neural network had as input. As a matter of fact, the air quality

[1] http://dati.arpa.puglia.it/dataset/meteo-nrt: weather dataset, last visited 23/01/2019.

[2] http://dati.arpa.puglia.it/dataset/aria-nrt: air dataset, last visited 23/01/2019.

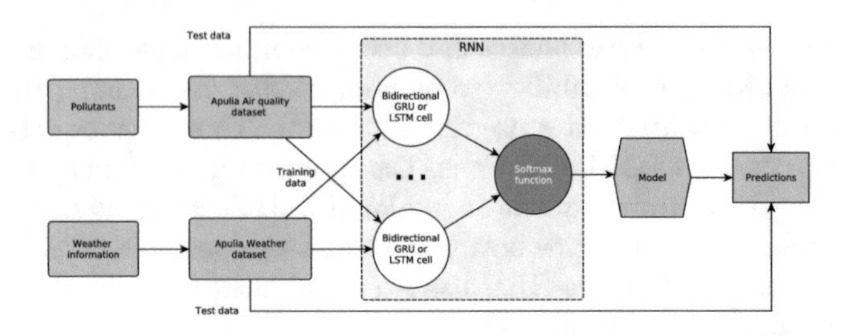

Fig. 1. The RNN model training and test schema.

station reports many pollutants value: PM_{10}, $PM_{2.5}$, NO_2, O_3, CO. However, among the available pollutants information, the Europe considers mandatory for the calculation of the CAQI only the NO_2, PM_{10}, O_3, while the $PM_{2.5}$, CO and SO_2 are optional [15]. To achieve a better accuracy in the predictions, in the proposed work, only the mandatory air pollutants are used. The workflow adopted to train the two RNN is shown in Fig. 1. The first step consists in the normalization and synchronization of all the data coming from the two datasets. As a matter of fact, both of them are near-real-time with a new acquisition each 10 min, however, coming from different type of stations they are not always perfectly synced. We conducted two different tests to prove that it is possible to reach an high accuracy in the prediction of the CAQI level, and that the RNN can be trained to forecast the CAQI level using weather and air information from different stations. The first considers the data ordered by time splitting it into two parts: the first one is used to train the RNN, while the second is used to test the network once the RNN model has been trained. The training process takes uses all the available days (from 2012 to 2017), then considers as test 5 random sequential days. In the second test it is asked to the RNN to make a "blind prediction". Firstly it has been randomly selected an air station and then it has been trained the RNN using the data coming from the others in a specific range (which by default has been set to 20 km). By calculating the angle between the selected station and the others, and analysing the weather conditions (including the wind intensity and direction), the RNN has been trained to predict the CAQI level. With the knowledge of the weather conditions of the previous days, the typical behaviour of the wind direction, the time of the year and the overall value of the pollutants, the RNN tries to estimate the value of each pollutant and the band of the CAQI. Both the RNNs have been modeled with three layers: one representing the input from the dataset, one inner layer (hidden) with 8 neuron cells (LSTM or GRU) and one as output as shown in Fig. 1.

4 Results

The two described dataset have been merged and fed to the RNN. In the first test performed a total of 1000 entries (almost the 80% of the dataset) have been used as training set, while 252 (almost the 20%) have been used as test

Table 2. The results (band error) of different models for daily average forecasting values of Air Quality: when the training and test samples are randomly selected (a); when the model has never seen data of predicting station (b).

Model	Forecasting	Bands	N. of days (accuracy) (a)	N. of days (accuracy) (b)
LSTM model	+1 day	±1	248(98.8%)	176(78.6%)
		±2	2(0.8%)	41(18.3%)
		±3	1(0.4%)	7(3.1%)
	+2 days	±1	246(98%)	177(79%)
		±2	4(1.6%)	44(19.7%)
		±3	1(0.4%)	3(1.3%)
	+3 days	±1	251(100%)	169(75.5%)
		±2	0(0%)	51(22.8%)
		±3	0(0%)	4(1.7%)
	+4 days	±1	247(98.4%)	175(78.1%)
		±2	4(1.6%)	45(20%)
		±3	0(0%)	4(1.8%)
	+5 days	±1	250(99.6%)	165(73.7%)
		±2	1(0.4%)	52(23.2%)
		±3	0(0%)	7(3.1%)
GRU model	+1 day	±1	247(98.4%)	170(75.9%)
		±2	4(1.6%)	44(19.6%)
		±3	0(0%)	9(4%)
	+2 days	±1	244(97.2%)	163(72.8%)
		±2	7(2.8%)	58(25.9%)
		±3	0(0%)	3(1.3%)
	+3 days	±1	249(99.2%)	159(71%)
		±2	2(0.8%)	57(25.5%)
		±3	0(0%)	7(3.1%)
	+4 days	±1	249(99.2%)	158(70.5%)
		±2	2(0.8%)	55(24.6%)
		±3	0(0%)	11(4.9%)
	+5 days	±1	251(100%)	157(70%)
		±2	0(0%)	58(25.9%)
		±3	0(0%)	8(3.6%)

set. To avoid the overfitting and the random weight initialization problem, we have repeated the training 20 times randomly choosing the training and testing data. The "blind prediction" test has a different configuration. It uses 1028

(82% of the dataset) entries as training and 224 (18% of the dataset) as test samples. The average of the final results of all trials are reported in Table 2. In the first column (Model) it is reported the model of the RNN, while the second one (Forecasting) represents the number of days in advance that the RNN have to predict. The third column (Bands) shows the CAQI value that the RNN is predicting (used like "bins" by the neural network). The last columns show how many days have been correctly predicted by the RNN when the test is performed on random data taken from the dataset or considering the blind test (respectively column "a" and column "b"). Results are excellent when it is asked the RNN to predict the CAQI level using the test data coming from the dataset. As a matter of fact, both the models are able to correctly predict with an accuracy close to the 100%. When the test data are hidden to the RNN, the accuracy in the predictions of both the models decrease. Results show that the LSTM model performs better than the GRU model trying to forecast the CAQI the band 1 one day ahead, while the opposite happens for the other bands.

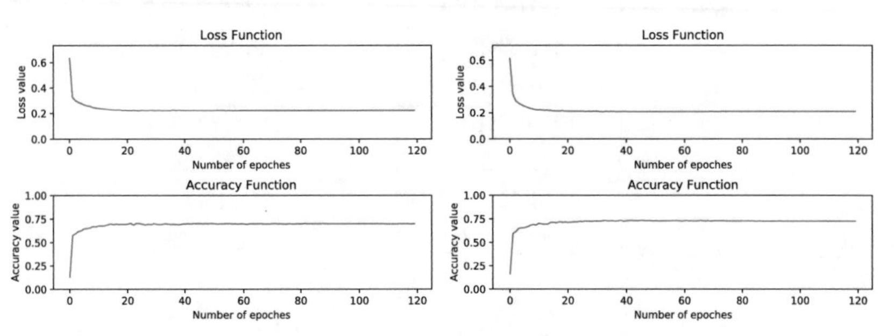

(a) The loss function and the accuracy function for the LSTM network

(b) The loss function and the accuracy function for the GRU network

Fig. 2. Both the LSTM and the GRU RNN have almost the same behaviour during the training process

In Fig. 2 are shown the loss functions for both the models during the training phase of the blind prediction per epoch. Despite the number of epochs used to train the RNN, and the accuracy reached on the test data, both the loss function and the accuracy for both the LSTM and the GRU exhibit the same behaviour.

5 Conclusions and Future Works

In this paper we have faced the problem of predicting the air pollution using the RNN. Using two different datasets about the weather information and the air quality of the Apulia region, we have trained two different RNN models using the LSTM an the GRU neuron cells respectively. We have adopted this kind of network to exploit both their ability to back-propagate the error, and their classification performances on dataset involving a time series. We have used only the CAQI mandatory pollutants and trained the RNN to predict the CAQI band for five days ahead. Two different types of experiments have been conducted: the former using as test some random data from the datasets, the latter letting the RNN to blindly predict the behaviour of a specific area knowing the behaviour of the neighbour one. Results shows that the RNN is able to predict with a very high accuracy the CAQI level of random days while the "blind prediction" results are promising.

To improve the current solution a more accurate dataset is needed. As a matter of fact, many entries from the dataset used in both the experiment were missing, thus have not been considered. To prove and tune the proposed "blind prediction" algorithm to install the model on the different air stations is needed.

Acknowledgment. The present study was developed and granted in the framework of the project: "SeVaRA" (European Community, Minister of the Economic Development, Apulia Region, BURP n. 1883 of the 24/10/2018, Id:2NQR592).

References

1. Anderson, J.O., Thundiyil, J.G., Stolbach, A.: Clearing the air: a review of the effects of particulate matter air pollution on human health. J. Med. Toxicol. **8**(2), 166–175 (2012)
2. Bevilacqua, V., Cassano, F., Mininno, E., Iacca, G.: Optimizing feed-forward neural network topology by multi-objective evolutionary algorithms: a comparative study on biomedical datasets. In: Italian Workshop on Artificial Life and Evolutionary Computation, pp. 53–64. Springer, Cham (2015)
3. Niska, H., Hiltunen, T., Karppinen, A., Ruuskanen, J., Kolehmainen, M.: Evolving the neural network model for forecasting air pollution time series. Eng. Appl. Artif. Intell. **17**(2), 159–167 (2004)
4. Feng, X., Li, Q., Zhu, Y., Hou, J., Jin, L., Wang, J.: Artificial neural networks forecasting of $PM_{2.5}$ pollution using air mass trajectory based geographic model and wavelet transformation. Atmos. Environ. **107**, 118–128 (2015)
5. Che, Z., Purushotham, S., Cho, K., Sontag, D., Liu, Y.: Recurrent neural networks for multivariate time series with missing values. Sci. Rep. **8**(1), 6085 (2018)
6. Huang, C.-J., Kuo, P.-H.: A deep CNN-LSTM model for particulate matter ($PM_{2.5}$) forecasting in smart cities. Sensors **18**(7), 2220 (2018)
7. Stevens, M., D'Hondt, E.: Crowdsourcing of pollution data using smartphones. In: Workshop on Ubiquitous Crowdsourcing (2010)
8. Adams, M.D., Kanaroglou, P.S.: Mapping real-time air pollution health risk for environmental management: combining mobile and stationary air pollution monitoring with neural network models. J. Environ. Manage. **168**, 133–141 (2016)
9. Sivacoumar, R., Bhanarkar, A., Goyal, S., Gadkari, S., Aggarwal, A.: Air pollution modeling for an industrial complex and model performance evaluation. Environ. Pollut. **111**(3), 471–477 (2001)
10. To, T., Shen, S., Atenafu, E.G., Guan, J., McLimont, S., Stocks, B., Licskai, C.: The air quality health index and asthma morbidity: a population-based study. Environ. Health Perspect. **121**(1), 46–52 (2012)
11. Cheng, W.-L., Chen, Y.-S., Zhang, J., Lyons, T., Pai, J.-L., Chang, S.-H.: Comparison of the revised air quality index with the PSI and AQI indices. Sci. Total Environ. **382**(2–3), 191–198 (2007)

12. Sowlat, M.H., Gharibi, H., Yunesian, M., Mahmoudi, M.T., Lotfi, S.: A novel, fuzzy-based air quality index (FAQI) for air quality assessment. Atmos. Environ. **45**(12), 2050–2059 (2011)
13. Caivano, D., Cassano, F., Fogli, D., Lanzilotti, R., Piccinno, A.: We@ home: a gamified application for collaboratively managing a smart home. In: International Symposium on Ambient Intelligence, pp. 79–86, Springer, Cham (2017)
14. Chung, J., Gulcehre, C., Cho, K., Bengio, Y.: Empirical evaluation of gated recurrent neural networks on sequence modeling. arXiv preprint arXiv:1412.3555 (2014)
15. World Health Organization: Air quality guidelines for Europe (2000)

Experiences in Context Aware-Services

Ichiro Satoh(✉)

National Institute of Informatics,
2-1-2 Hitotsubashi, Chiyoda-ku, Tokyo 101-8430, Japan
ichiro@nii.ac.jp

Abstract. Context-aware services are one of the most typical services in ambient intelligent environments. Although there have been many academic projects on context-aware services, most of them were evaluated in laboratory-level or small-scale experiments. Context-aware services in real spaces suffer from a variety of problems. This paper addresses such problems. It describes experiences learned from our experiments on context-aware services in real public spaces, e.g., museums, and proposes solutions to them. Most of the problems presented in this paper are still common to other context-aware services.

Keywords: Experience · Ambient intelligence · Evaluation

1 Introduction

Context-aware services are one of the most typical applications of ambient intelligence. There have been a huge number of academic experiments on context-aware services, but most of them were laboratory-level experiments with small number of participants. We have not only been explored several middleware systems for context-aware services with sensing systems, but also provided real context-aware services for real users in the real world, e.g., museums and stores, because context-aware services essentially depend on the real worlds. We have gained knowledge from our experiences. This paper focuses on describing lessons learned from our experiments on context-aware services, in particular location-aware services, in the real world, rather than our proposed systems. Since these lessons are common to other context-aware services, we believe that they will be useful to other systems.

The remainder of this paper is organized as follows. In Sect. 2 we describes our experiment. Section 3 presents our lessons learned from the experiment. Section 4 surveys related work and Sect. 5 provides a summary.

2 Experiment

This paper addresses one of our experiments as an example to discuss problems in providing context-aware services with real users in the real world. The

P. Novais et al. (Eds.): ISAmI 2019, AISC 1006, pp. 45–53, 2020.
https://doi.org/10.1007/978-3-030-24097-4_6

example was designed and operated as a location-aware user-assistant system in a science museum at the Museum of Nature and Human Activities in Hyogo, Japan [5,6]. We conducted the experiment over one month. Most visitors in the museum lacked sufficient knowledge about exhibits and they needed supplemental annotations on these. As their knowledge and experiences were varied, they may have become puzzled (or bored) if the annotations provided to them were beyond (or beneath) their knowledge or interest. To solve this problem, we constructed and provided a user/location-aware system to assist visitors at the museum. The experiment was carried out at several spots in front of specimens of stuffed animals, e.g., a bear, deer, racoon dog, and wild boar. Each spot could provide five different pieces of animation-based annotative content about the animals, e.g., their ethology, footprints, feeding, habitats, and features, and each had a display and a Spider's active RFID reader with a coverage range that almost corresponded to the space, as seen in the left of Fig. 1.

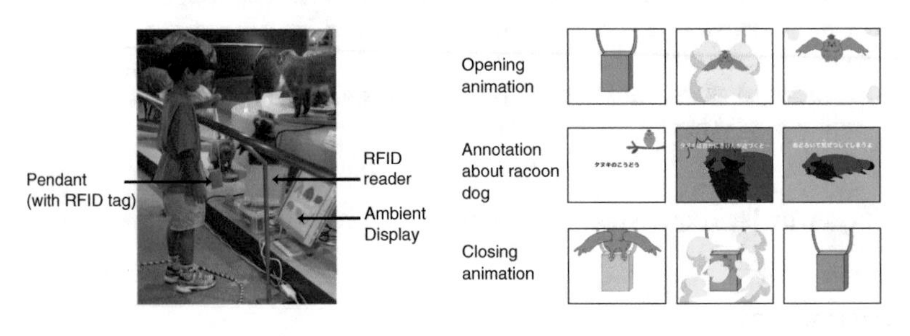

Fig. 1. Spot at Museum of Nature and Human Activities in Hyogo (Left) and Opening animation, annotation animation, and closing (Right)

When a visitor entered a spot with the specimen of a racoon dog, his/her visual agent migrated from his/her pendant to the display located in the spot. As we can see from the right of Fig. 1, a visual agent tied to the orange pendant plays the opening animation to inform that its target is a visitor with an orange pendant, where the animation displays the agent's character appearing as an orange pendant. The experiment offered visitors animation-based annotative content about the animal in front of them so that they could learn about it while observing the corresponding specimen.

3 Lessons Learned from Experiment

This section describes several problems with which we were confronted in the experiment and discusses our solutions to them.

3.1 Missing Context-Aware Services

Most existing experiments on context-aware services explicitly or implicitly assume that users know where, when, who, what, and how to receive the services. However, in fact, users often fail to recognize their context-aware services.

Most visitors at our early experiment in a museum missed location-aware services because they did not know where they could receive the services. They also wandered around the exhibition room and haphazardly entered enter certain places where they could receive the services.

- While they were not at the right places, they could not know where and what services were provided to them because the services were available in certain contexts, e.g., particular locations and users.
- Several users were surprised when they stood at places because they were suddenly provided with services, even when they had knowledge about the availability of location-aware services at some places.

It was difficult for users to know where, when, who, what, and how context-aware services were provided, because context-aware services, e.g., location-aware, time-aware, and user-aware, were only provided while the current contexts, e.g., locations, times, and users, satisfied the conditions of the services.

To notify the spots that context-aware services are provided, we provided a simple solution to this problem after several attempts. We placed visual markers on the floor in front of places which location-aware services were provided to visitors. Figure 2 shows two kinds of markers we used. The first was a printing sheet and quoits. After doing this, all visitors could find the places. This seemed to be a simple but effective approach to solving the problem.

Fig. 2. (a) and (b) Markers on location-aware services and (c) two or more visitors at a spot

3.2 Supports to Multiple Users

As discussed above, the experiment used stationary terminals instead of mobile terminals to enable visitors to pay their attention to seeing exhibitions as much as possible. Each stationary terminal may be shared by more than one visitor, whereas each mobile terminal can be carried and used by an individual visitor or a group of visitors. Stationary terminal have a problem that portable terminals do not have. The IoT for urban should support such stationary terminals, which may be shared among multiple users.

In the experiment, we made five pendants with RFID tags so that at most five individual visitors or groups could participate at the same time. Children tended to gather the same spot as shown Fig. 2. Our system was equipped with a queuing mechanism for exclusively executing visual agents for multiple simultaneous visitors at each of the spots. When two visitors entered the same spot, the system activated the agent of either of them and displayed on the screen of a stationary terminal in the spot in the order in which they entered. However, they could not identify their agents or others, or incorrectly identify the agents of others as theirs. There were many visitors in museums so that we could not cope with conflicts caused by multiple users.

To solve this problem, the experiment was designed to enable visitors to imagine that their agents, which were virtual owls, were within their pendant. Each visitor has a colored pendant including RFID tags, where these pendants are green, orange, blue, yellow, or red. He/she has a visual agent with color corresponding to the color of the pendant attached to the agent, because visitors could distinguish between their agents and others' agents through their pendants' colors. This may seem to be naive but it effectively enables visitors to readily identify their agents. This is available with other context-aware services in the IoT.

3.3 Supports to Human Errors

Several existing research projects implicitly assume that users always do as their system hope. However, many users do not this. When we provided navigation services from the terminals in the experiment, some of the visitors did not follow the navigation services. They have two different types of human errors: users did not notice information from the navigation services and did not follow the navigation; or users noticed the information but their performance can be deficient. Human error is inevitable because no one is perfect. It may not be possible to completely prevent human error, but is very much possible to minimize or mitigate it.

We have two approaches to solving the first error: the system provide users with the same information again; or enables users to know that they missed the information that they should know. To solve the second error, we have two approaches: the system informs users about their errors and then leave modification to them; or provides information for them to recovery themselves, even when they do not noticed their errors. However, users who missed visual information may also miss visual warning information, because they paid their visual attention to other targets.

Our experiment notified visitors about their errors, when the system detected the errors through sensing systems. To solve the first errors, it provided visitors with audio information in addition to visual information, when it detected that they did not follow its navigation.

3.4 Handling of Sensing Errors

Since there are not any perfect sensing systems, sensing errors are inevitable. Although there is no silver bullet to solve this problem, there are some approaches to avoiding the problem. One of them is to find errors of some sensing systems by using other sensing systems in combination. However, this approach is often expensive. Our experiment selected location sensing systems according to the tendency of their potential errors. Although a sensing system can measure the locations of target objects in detail, if it tends to have a large deviation in the locations, it should not be used.

This experiment used proximity-based positioning rather than lateration-based one. This is because the deviation of locations measured by the former can be smaller than that by the latter. In our proximity-based sensing system radio-receivers might miss radio signals issued from RFID tags within the coverage areas of the signals, but they could not receive signals issued from RFID tags far from the receivers.

3.5 Rapid/Easy Installation and Deployment

Many academic projects have provided services in their laboratories. However, existing real spaces are already been being used by people for their own purposes. Also, constructors of context-aware services in real spaces may not have any professional knowledge of context-awareness. We also tend to forget about logistics for context-aware systems. The deployment and configuration of context-aware services in the real spaces have time limitations. We had to deploy, install, and configure the entire system, including location-sensing systems and servers, by us within one day in the experiment, i.e, a day the museum was closed, without conducting any preparatory experiments in the exhibition room.

Fig. 3. Invisible installation (left) and invisible installation (right)

Our system was structurally constructed as a set of building blocks and easily assembled in the target space for a few hours to overcome these limitations. It supported an autonomic configuration for automatically detecting sensors and computers in a peer-to-peer manner. Our systems were encapsulated in portable containers as building blocks (the left of Fig. 3), which were ISO-standardized stackable boxes for logistics. The containers could directly be used in carrying and operating the system. Since they were stackable, they could be easily and safely be carried by a logistics company.[1]

3.6 Adaptation to Legacy Spaces

Many academic projects explicitly or implicitly assume that their context-aware services are provided in brand-new spaces, e.g., buildings, floors, and rooms. There is no space for computers and sensors in real spaces, including cities. No power-lines and networks may be available. Furthermore, there may be no space for any management systems in experiments. These problems are common. For example, city-wide context-aware services are often required to be monitored and managed at target spaces, which have no terminals for management tasks. In an experiment in the National Science Museum, the museum asked us to hide all devices from the view of visitors (the right of Fig. 3). It is more difficult to make legacy spaces than to build new smart spaces because of their constraints. The deployment of context-aware services needs to make existing spaces smart without losing any utility of the original spaces. Several museums have also required all devices to be invisible to visitors.

Terminal for monitoring the positions of
visitors and customizing agents

Fig. 4. Portable management terminal (left) and monitoring tool (right)

Our system enabled administrators to operate and monitor the system through portable terminals instead of any stationary terminals. It provided Web-based APIs to create, control, and terminate agents in the current implementation. Therefore, an operator could create and customize agents through a Web browser running on his/her (portable) computer. The left of Fig. 4 shows our GUI interface for binding agents to users, running on a portable terminal equipped with a WiFi interface (Apple iPod Touch). Each runtime system provides an HTTP server to be monitored from the external systems for administration reason.

[1] They could be carried inexpensively because they were standardized and stackable.

3.7 Interactive vs. Non-interactive Services

Context-aware services in public spaces can be classified into two modes.

- *Non-interactive mode:* Context-aware systems often have the ability to automatically adapt themselves to users' needs without their participation. Although this mode may not be able to support rich services, users do not have to explicitly operate the systems.
- *Interactive mode:* There are a variety of possible services provided by systems that support this mode, but users must explicitly operate the systems through devices. Although there has been a great deal of work on computer usability, it is still difficult for children, elderly, and handicapped people to manipulate such devices.

We should select the two modes according to users and services.

Several visitors were annoyed with interactive services because they only wanted to know about annotation on the exhibits. Furthermore, interactive services often tend to lower the extent of visitors' learning experiences in museum, because they tend to be preoccupied with interactions.

A solution to this problem makes annotation services context-aware. However, it is difficult to detect users' contexts by using sensing systems. Instead, we divided our annotation services into first and second halves. The first provided non-interactive annotation services and the second interactive ones. We enabled visitors to receive annotations about the exhibits in the first and move to other exhibits without being provided with the second.

3.8 Context-Aware Services as Extra Roles Rather Than Leading Roles

Many researchers have had experiences to evaluate context-aware services. Nevertheless, most of them intended to evaluate the services themselves rather than their utility or effectiveness for users to achieve their own goals. The main purpose of context-aware services is to assist users' activities. Context-aware services are just methods instead of any goals. In fact, one goal of science museums is to provide visitors with experiences to visitors that will enhance their knowledge of science from exhibitions, but not with devices. Several existing projects have used context-aware visual agents, e.g., characters or avatars, on terminals to attract the interest of children. Therefore, such services should play as extra roles instead of leading roles. Several visitors tended to focus their attention on visual agents rather than exhibits.

Our solution is to use context-aware services as extra roles rather than leading roles. Context-aware services should be designed to play supporting or extra roles because exhibit are meant to play leading roles in museums. Even when exhibitions assumed the use of our context-aware services, we designed them to enable visitors to enjoy exhibitions without context-aware services. The experiment initially provided animations with visual agents, Later in the experiment, we had to abandon the animations, because some visitors tended to focus attentions on visual agents rather than exhibits.

4 Related Work

As we discussed in the first section, a few academic projects evaluated their services with real users in real spaces. Many academic attempts on context-aware services have been developed to the prototype stage and tested in laboratory-based or short-term experiments with professional administrators. They have also been designed in an ad-hoc manner to provide specific single services in particular spaces, i.e., research laboratories and buildings. As a result, they have not been suitable for public spaces or for applications that they were not initially designed to support. In fact, there is a gap between laboratory-level or prototype-level systems and practical systems.

This section only discusses some related work that has provided real applications for real users in museums, because the experiment discussed in this paper aimed at context-aware services in a science museum. One of the most typical approaches in public museums has been to provide visitors with audio annotations from portable audio players. These have required end-users to carry players and explicitly input numbers attached to the exhibits in front of them if they wanted to listen to audio annotations about the exhibits. Many academic projects have provided portable multimedia terminals or PDAs to visitors. These have enabled visitors to interactively view and operate annotated information displayed on the screens of their terminals, e.g., the Electronic Guidebook, [2], the Museum Project [1], the Hippie system [4], ImogI [3], and Rememberer [2]. They have assumed that visitors are carrying portable terminals, e.g., PDAs and smart phones and they have explicitly input the identifiers of their positions or nearby exhibits by using user interface devices, e.g., buttons, mice, or the touch panels of terminals. However, these existing work has only involved short-term experiments with few participants. They did not described such problems with context-aware services in the real world as discussed in this paper.

5 Conclusion

Although there have been many academic projects on context-aware services, many potential problems in the real world are not well-known. Therefore, this paper described our lesson learned from experiences on context-aware services. This paper described several problems and our solutions to them. We believe these problems will become common in others' projects on context-aware services.

References

1. Ciavarella, C., Paterno, F.: The design of a handheld, location-aware guide for indoor environments. Pers. Ubiquit. Comput. **8**(2), 82–91 (2004)
2. Fleck, M., Frid, M., Kindberg, T., Rajani, R., O'BrienStrain, E., Spasojevic, M.: From informing to remembering: deploying a ubiquitous system in an interactive science museum. IEEE Pervasive Comput. **1**(2), 13–21 (2002)

3. Luyten, K., Coninx, K.: ImogI: take control over a context-aware electronic mobile guide for museums. In: Workshop on HCI in Mobile Guides, in Conjunction with 6th International Conference on Human Computer Interaction with Mobile Devices and Services (2004)
4. Oppermann, R., Specht, M.: A context-sensitive nomadic exhibition guide. In: Proceedings Symposium on Handheld and Ubiquitous Computing (HUC 2000). LNCS, vol. 1927, pp. 127–142. Springer, September 2000
5. Satoh, I.: Context-aware agents to guide visitors in museums. In: Proceedings of 8th International Conference Intelligent Virtual Agents. LNCS, vol. 5208, pp. 441–455. Springer (2008)
6. Satoh, I.: Building context-aware services from non-context-aware services. In: Proceedings of 3rd International Symposium on Ambient Intelligence (ISAmI 2012). Advances in Intelligent and Soft Computing, vol. 153, pp. 59–66. Springer (2012)

Gesture Control System for Industry 4.0 Human-Robot Interaction – A Usability Test

Luis Roda-Sanchez[1(✉)], Teresa Olivares[1,2], Arturo S. García[1,2], Celia Garrido-Hidalgo[1], and Antonio Fernández-Caballero[1,2]

[1] Instituto de Investigación en Informática de Albacete, Universidad de Castilla-La Mancha, 02071 Albacete, Spain
luis.roda@alu.uclm.es, {Teresa.Olivares,ArturoSimon.Garcia,Celia.Garrido,Antonio.Fdez}@uclm.es
[2] Departamento de Sistemas Informáticos, Universidad de Castilla-La Mancha, 02071 Albacete, Spain

Abstract. The Industry 4.0 paradigm pursues improvements in production rate, flexibility, efficiency, quality, among others, through the use of technologies like Internet of Things (IoT), ambient intelligence and collaborative robots. Robots developing precision tasks, works in hazardous environments or movements of heavy parts, autonomously or in cooperation with workers, offer great advantages. Although collaboration provides great benefits, these technologies should be appropriate for all kind of workers, independently of their technical skills. If this problem is not addressed properly, irruption of robots could lead to social instability and/or rejection of useful advances. In this work, a gesture control system based on wearables oriented to Industry 4.0 robots is tested with real users to validate a novel gesture control system as an intuitive tool.

Keywords: IoT · Industry 4.0 · Gesture control · Human-robot interaction · Ambient intelligence

1 Introduction

The arrival of Industry 4.0 is a big step forward for many companies. Indeed, Industry 4.0 pursues improvements in production rate, flexibility, efficiency, quality, among others, through the smart use of technologies like IoT, ambient intelligence, collaborative robotics and agent-based software engineering (e.g. [1,2]). It involves major changes to some traditional and manual processes such as movement of heavy and/or dangerous materials, precision tasks or works in harmful environments. However, this digitization could produce a social instability as well, since there is an imbalance among workers' technological skills. This will probably cause serious problems in the near future and lead to the rejection of technological breakthroughs. For this reason, the invention of intuitive and comfortable tools for workers will avoid social and health problems.

P. Novais et al. (Eds.): ISAmI 2019, AISC 1006, pp. 54–61, 2020.
https://doi.org/10.1007/978-3-030-24097-4_7

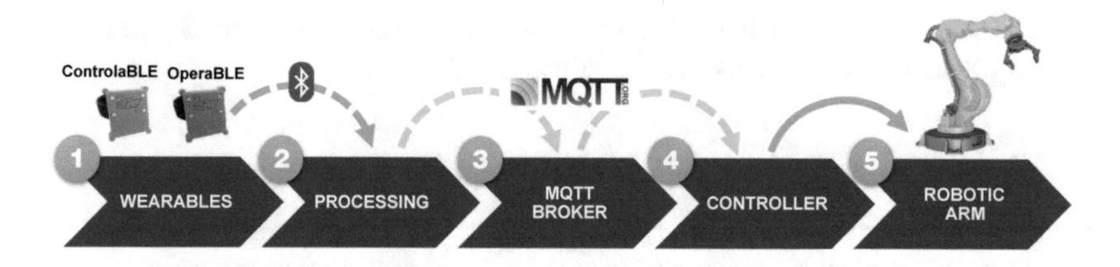

Fig. 1. Diagram of the gesture control system.

There are already some related contributions oriented to industrial applications. Recently, a machine learning approach has been proposed for industrial prognosis [3]. The authors review different relevant models to achieve failure prediction. In another work [4], technologies available for obtaining data from workers' user experience are explored to improve their well-being and security. Even more closely related to robots and gesture control, a control system based on computer vision (Microsoft Kinect) using fuzzy logic has been introduced [5].

In our previous research [7,8] and [6] we have presented wearables to improve workers' conditions; a complete deployment of a heterogeneous network to fulfill requirements of Industry 4.0 scenarios and a review of wireless sensor networks. In this work, a usability test of our recently developed gesture control system is carried out, performed by a diverse group of users. The selection of candidates was focused on several aspects such as age, field of knowledge and use of technology to obtain a wide range of people and not so much a great number. First, users performed several tests using the gesture control to move a robotic arm and, after that, they completed a questionnaire adapted from the USE Questionnaire [9].

2 System Description

Figure 1 shows a simple descriptive diagram of the proposed system. There are four stages within the system. The first is composed of two wearables put on the users' arms to control the robotic arm. OperaBLE is the name of the wearable that acquires movement information aimed at performing several robotic arm actions. The other wearable, called ControlaBLE, sends control commands to the processing board. Table 1 shows recorded commands for each wearable. The users control the robot to test the degree of naturalness through these commands.

The processing board conforms the second stage, where the movement is processed to obtain an action command. Movement recognition is performed by LoMoCA [7], which is our proper low-frequency movement characterization algorithm, created to reduce the processing time in low-power devices. An MQTT (Message Queuing Telemetry Transport) broker is also included in this device to manage communications with the controller board. The controller board is attached to the robot. It consists of a Single Board Computer (SBC) that manages the reception of commands via MQTT from the processing board.

The SBC is attached to a controller board to send signals to 6 joints of the robot to perform several predefined movements.

Table 1. Commands available through each wearable.

OperaBLE				
Forward	Backward	Left	Right	Pick/Drop
Right profile				
① ②	② ①	② ①	① ②	③ ① ②
ControlaBLE				
Inactive	Low	Medium	High	Emergency
Left profile				

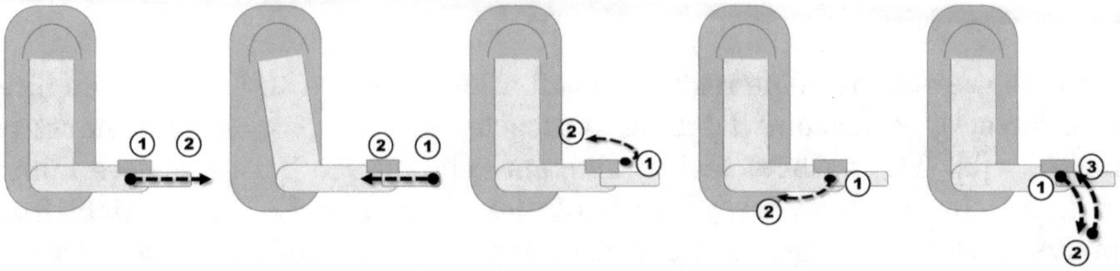

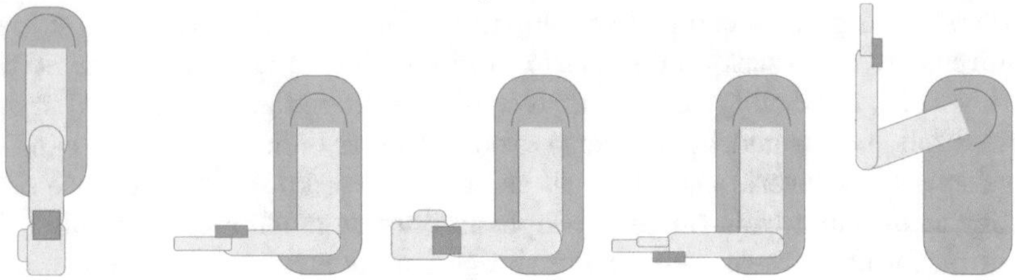

3 Methodology and Validation

In regard to the testing process, there are several questionnaires that are useful to unveil the degree of naturalness perceived by the users of the control system. However, in this case, the chosen set of questions was adapted from the so-called USE questionnaire [9], since it seems adjusted to assess the prototype. The block of questions about usefulness was omitted due to its inadequacy for evaluating prototypes, since it is thought to test devices and systems that are widely used.

3.1 Experiment Setup

The experiment setup was developed following several steps in order to achieve a better adjustment of the form and tasks to perform by all kind of user. First, a small group of 5 people without any knowledge about the operation of the system was selected to obtain a first feedback from users. At this point, some errors in the way of asking were detected. Therefore, this first step was useful to find ambiguities in the form, which could lead to misunderstandings. Several questions were revised, achieving a second version of the form.

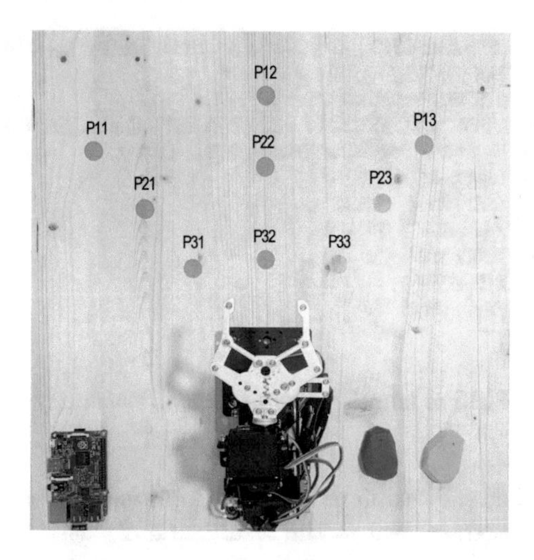

Fig. 2. Work space used to perform the usability test.

Then, the experiment was carried out again, in this case choosing a group of 3 experts in the field. The majority of comments provided were focused on the manner of performing the experiment to achieve reliable data.

Once the questionnaire and the experiment were properly adapted, there were several aspects which had to be considered in the final version tested by 15 users. The purpose of this study was to evaluate to which extent users perceived natural interaction with the control system [10]. For this reason, patterns recorded for each movement remained unchanged to avoid disparities during the process, which could again lead to misleading conclusions. In future works, it would be interesting to perform the same study, but changing this parameter to observe how much it affects the assessment.

The proposed tasks for testing the operation of the control system began with an adaptation period where users practiced to get used to the manner of performing the movements (adjusted to the patterns). The work space used to carry out this test is offered in Fig. 2. After few minutes, the users had to move the robot from position P32 to P11 and back again to P32. Finally, a more complex task had to be done; Users should pick up an object placed in P21, carry it through the matrix from P21 to P23, drop it and activate the emergency state.

3.2 Participants and Questionnaire

The selection of users required to evaluate the system included different types of people. Hence, many aspects were considered to achieve a reasonable sample of users. The choice focused on the following aspects, which were thought to reach significant conclusions: age, area of knowledge and use of technology.

Therefore, a great number of users was not the aim, as we intended a heterogeneity of people to perform the test to identify relationships between the factors.

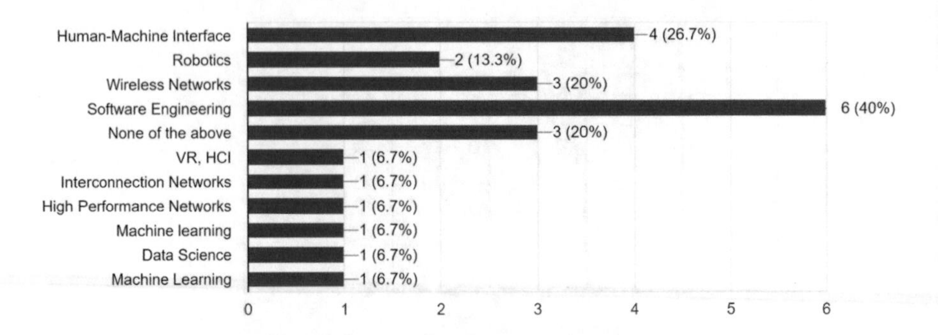

Fig. 3. Field of knowledge of users with university studies.

In Fig. 3 a graph which shows different fields of knowledge of users (holding university studies) is depicted. In Sect. 4, the results and discussion related to the connections observed, identified weaknesses and answers provided by the 15 users that performed the test will be shown.

Regarding age, users were between 22 and 55 years old, which is a wide sample enabling to observe possible changes in the understanding and performance of the experiment. Users also answered a question concerning use of technology in their daily life. The majority of people use technology daily (80%). However, 13.3% are occasional users and, finally, 6.7% do not use technology. It is necessary to clarify who we have considered to be a technological user. In this study, a complete use of technology would be people that are really skilled using computers and smartphones, occasional use if they employ technology to develop minor works. Eventually, a non-technological user in this case would be someone who has difficulties to accomplish simple tasks with both devices.

The final structure of the questionnaire is composed of previous questions to gather data from users and three blocks from the USE questionnaire: ease of use, ease of learning and satisfaction. Each question can be marked from 1 to 5 points, where the maximum value corresponds to the best possible mark.

4 Results and Discussion

In this section, several graphs related to the experiment carried out are shown. It is divided in three main blocks, following the proper structure of the questionnaire. Therefore, each part will show the explanation and discussion separately.

Ease of Use. The aim of this block of questions was to know the perception of users about the gesture control system's naturalness and simplicity of use. Due to the breadth of this block of questions, the most significant graphs have been selected to focus on the information provided by the users. Figure 4 details the results of asking if the control system is considered easy to use.

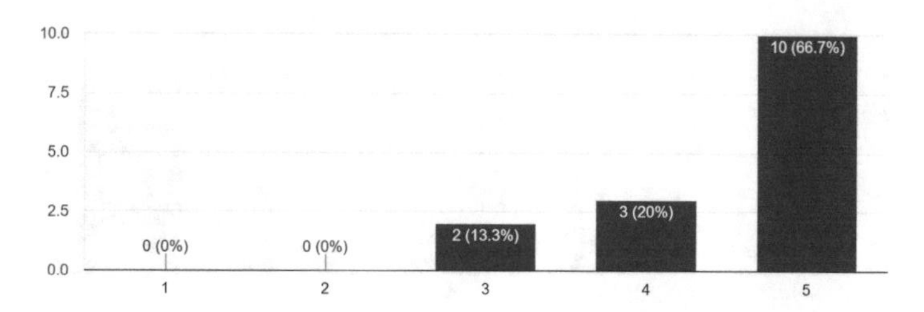

Fig. 4. Answers to "The control system is easy to use".

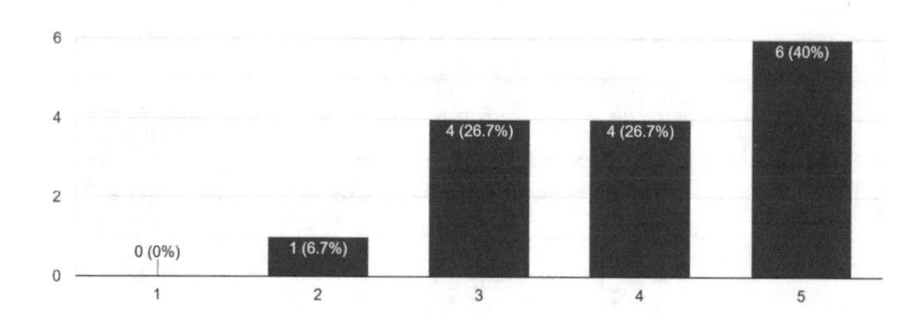

Fig. 5. Answers to "Using the control system is effortless".

It seems clear that the control system is easy to use for different kinds of people. In fact, observations made during the experiment show that the answer to ease of use does not depend on age, since there were no differences between young and old people. Nor does the type of studies of each user influence on the answers. Another interesting graph is shown in Fig. 5, demonstrating that this system's use implies some effort for many people.

Further comments and observations showed that the most difficult task was probably to coordinate both arms during the test. Gestures performed with OperaBLE are usually quite natural. However, operating ControlaBLE is not as intuitive as intended. Obviously, this fact causes a decrease in user-friendliness. This is a good starting point to improve the system in following studies.

Despite diversity of opinion in regard to the effort needed, almost all users thought that occasional and regular users would like this control system, selecting 2 points 6.7%, 4 points 20% and the highest score 73.3% in this question. Furthermore, to develop the test, no written instructions were provided by users, since it was not required for understanding the operation properly.

Ease of Learning. According to the questionnaire used to carry out the test, the next block of questions aims to assess the ease of learning of the gesture control system. Figure 6 shows a graph linked to the learning curve. In this case, near 90% users learned to use the control system quickly.

The previous graph throws similar answers to the given by the participants to the question "I easily remember how to use the control system". The results show a totally understandable system and users can retain the information effortlessly.

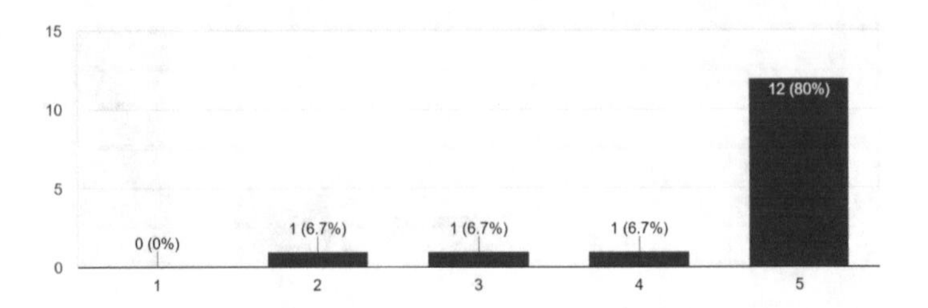

Fig. 6. Answers to "I learned to use the control system quickly".

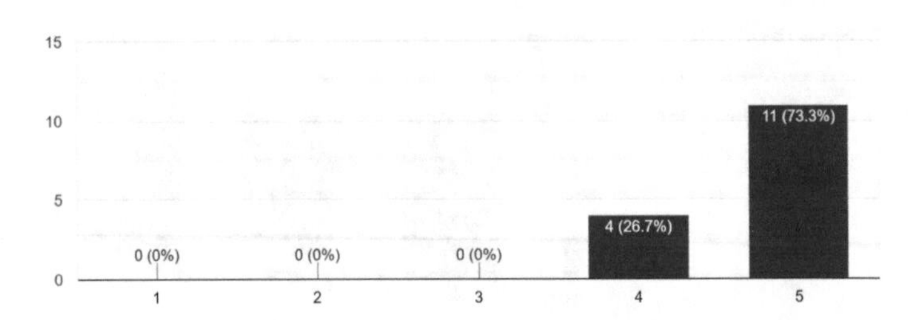

Fig. 7. Answers to "I would recommend the control system to workers".

Satisfaction. Users had as well to judge their general satisfaction in respect to the control system. As could be observed, the system to control the robotic arm is well received by users, since five of them gave four points in this question and ten the highest mark. The use of motion sensors such as accelerometers and gyroscopes [11], and the recognition achieved by LoMoCA, seems to be adequate for all users. In fact, Fig. 7 shows that users would recommend this control system to industry workers for developing non-programmed tasks. Therefore, it could be an appropriate gesture control system for future developments oriented to Industry 4.0 robots and new alternatives to human-robot interaction [12].

5 Conclusions

The results of performing this study with a diverse group of people show that the control system proposed is intuitive for a wide range of users. However, the operation of both wearables at the same time seems to be a bit confusing for some people. Therefore, this is an aspect to be considered in future works to reduce the effort perceived by some potential users.

The questions related to the learning process demonstrate that it is really easy to use the gesture control system in a very short period of time, even without written instructions. Moreover, a significant characteristic is that there is no difference in performance between men and women, young and old people, people with different studies and users with or without technological skills. This fact results in a substantial benefit to this type of control because it could contribute towards social balance, since all workers could use it and be part of the next industrial revolution.

Acknowledgements. This work was partially supported by Spanish Ministerio de Ciencia, Innovación y Universidades, Agencia Estatal de Investigación (AEI)/European Regional Development Fund (FEDER, UE) under DPI2016-80894-R grant.

References

1. Gascueña, J.M., Navarro, E., Fernández-Sotos, P., Fernández-Caballero, A., Pavón, J.: IDK and ICARO to develop multi-agent systems in support of ambient intelligence. J. Intell. Fuzzy Syst. **28**(1), 3–15 (2015)
2. Rivas-Casado, A., Martinez-Tomás, R., Fernández-Caballero, A.: Multi-agent system for knowledge-based event recognition and composition. Expert Syst. **28**(5), 488–501 (2011)
3. Diez-Olivan, A., Del Ser, J., Galar, D., Sierra, B.: Data fusion and machine learning for industrial prognosis: trends and perspectives towards Industry 4.0. Inf. Fusion **50**, 92–111 (2019)
4. Peruzzini, M., Grandi, F., Pellicciari, M.: Benchmarking of tools for user experience analysis in Industry 4.0. Procedia Manuf. **11**, 806–813 (2017)
5. Dajun, Z., Minghui, S., Fei, C., Chih-Min, L., Longzhi, Y., Changjing, S., Changle, Z.: Use of human gestures for controlling a mobile robot via adaptive CMAC network and fuzzy logic controller. Neurocomputing **282**, 218–231 (2018)
6. Hortelano, D., Olivares, T., Ruiz, M.C., Garrido-Hidalgo, C., López, V.: From sensor networks to Internet of Things. Bluetooth Low Energy, a standard for this evolution. Sensors **17**(2), 372 (2017)
7. Roda-Sanchez, L., Garrido-Hidalgo, C., Hortelano, D., Olivares, T., Ruiz, M.C.: OperaBLE: an IoT-based wearable to improve efficiency and smart worker care services in Industry 4.0. J. Sens. **2018**, 6272793 (2018)
8. Garrido-Hidalgo, C., Hortelano, D., Roda-Sanchez, L., Olivares, T., Ruiz, M.C., López, V.: IoT heterogeneous mesh network deployment for human-in-the-loop challenges towards a social and sustainable Industry 4.0. IEEE Access **6**, 28417–28437 (2018)
9. Lund, A.: Measuring usability with the USE questionnaire. Usability and User Experience Newsletter of the STC Usability SIG, vol. 8 (2001)
10. López-Jaquero, V., Montero, F., Molina, J.P., Fernández-Caballero, A., González, P.: Model-based design of adaptive user interfaces through connectors. In: Jorge, J.A., Jardim, N.N., Falcão e Cunha, J. (eds.) Interactive Systems - Design, Specification, and Verification, pp. 245–257. Springer, Heidelberg (2003)
11. Neto, P., Pires, J.N., Moreira, A.P.: Accelerometer-based control of an industrial robotic arm. In: The 18th IEEE International Symposium on Robot and Human Interactive Communication, vol. 11, pp. 1192–1197 (2009)
12. Mendes, N., Ferrer, J., Vitorino, J., Safeea, M., Neto, P.: Human behavior and hand gesture classification for smart human-robot interaction. Procedia Manuf. **11**, 91–98 (2017)

Attribute Grammar Applied to Human Activities Recognition in Intelligent Environments

Leandro O. Freitas(✉), Pedro Rangel Henriques, and Paulo Novais

ALGORITMI Center, University of Minho, Braga, Portugal
leanfrts@gmail.com

Abstract. Researches about context awareness have been growing in the past decades. The development of services that considers the context of users are getting popular and are gaining more functionalities, making them smarter. One of the most common features is the monitoring of activities through a diversity of sensors. Yet, this is still superficial monitoring where the devices lack information sharing. Intelligent environments aim the exchanging of information with the purpose of creating models that represent real-world situations. This paper describes the use of an attribute grammar in order to create a formal specification of situations in such domains. The problem of representation of human activities is tackled through a case study to demonstrate how attribute grammar can help the improvement of this process.

Keywords: Context-aware systems · Human activities recognition · Attribute grammar

1 Introduction

Identification of patterns behaviour of users is considered an obstacle for the consolidation of intelligent environments. Systems for these domains should be aware of the users' steps along the day in order to identify what and how they perform their daily actions. Monitoring activities allow the system to improve its capability of assisting users by adjusting the environment according to their preferences or even helping them to solve problems. These systems have multiple agents capable of processing context data and sharing it with each other, aiming to increase the level of its intelligence. Context can be defined as anything that can influence the behaviour of users [5].

One of the main problems when developing context-aware systems refers to the detection of environmental data, once this will be used as a premise for processing and understanding what happens in the domain. Thus, incomplete or lack of enough context data may compromise this process leading to uncertain situations, and consequently, the decision making by the system. Solutions for the real world should take into account different levels of uncertainty [11]. The

P. Novais et al. (Eds.): ISAmI 2019, AISC 1006, pp. 62–70, 2020.
https://doi.org/10.1007/978-3-030-24097-4_8

representation of it improves assertions about the domain's knowledge. This tends to minimize problems related to the design of the model and, consequently, the negative impact of misinterpretation of contexts [1].

Modelling human activities is not an easy task. It is necessary to develop means that allows the manipulation of context data and use it to create models that represent what happens in the environment. Besides that, it is necessary to validate this through a formal specification. Attribute Grammar (AG) [7] is a field of investigation that can be applied for the validation of human activities recognition in intelligent environments due to its formalism.

Considering this, the main goal of this work is to describe how AGs can be used to tackle this problem. It is presented how they can improve the correctness of context data analysis. For this, the paper is structured as follows: Sect. 2, describes concepts of AGs and how they can be applied to intelligent environments, once not much it is known about their relation. Section 3 presents an example of AG for human activities recognition aiming to demonstrate its usefulness. Related work is described in Sect. 4. The benefits of AGs applied to context-aware systems and final considerations of the paper are discussed in Sect. 5.

2 Attribute Grammars Applied to Intelligent Environments

Attribute Grammar (AG) is a wide field of investigation with several relevant contributions regarding the validation of programming languages. It introduces semantic rules into Context-Free Grammars, inserting meaning to the analysis of the grammar [7]. Still, according to [7], the semantics for the rules is achieved through the definition of attributes for non-terminal symbols by associating them with each production. This way, they are defined to manipulate attributes, which are associated with the grammar symbols.

The lexical and syntactical phases of processing precede the semantic. The lexical analysis has as the main goal to analyse the inputs and convert them into a set of terminal symbols. The result of this processing is used in the syntactical phase to generate an Abstract Syntax Tree (AST) [4]. These two first phases can be applied to intelligent environments once this kind of domain should be filled with sensors sending raw data to be analysed by a context-aware system. Then, the semantic analysis is performed with the goal of taking unevaluated attributes as input and returning evaluated ones.

Another characteristic of AGs is the possibility of manipulation of the attribute values at any node of the tree [4]. This feature is very important once in an intelligent environment the states of entities (attribute values in AGs) may change frequently and their values must be available to be used by the system at any time, with different purposes. A generic Abstract Syntax Tree of an intelligent environment is presented in Fig. 1.

Te terminal symbols, in **a**, represent the raw data sent by sensors from the intelligent environment, in **b**. It refers to any sort of data captured by the sensors

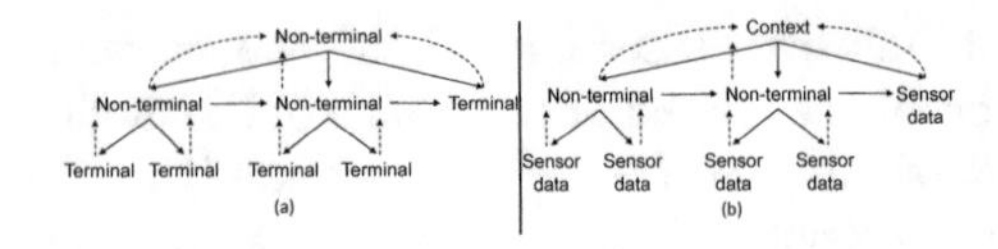

Fig. 1. **a** General structure of AST. **b** Generic AST for intelligent environments.

(e.g. humidity, level of lightness, localization, the presence of entities in a room). It will be used to characterize new situations (new context). In the general structure **a**, it will be characterized as been synthesized or inherited attributes and can be analyzed in different nodes of the tree. The entities represent the non-terminal symbols in an AG and may assume all the characteristics of them. They can be composed of several data properties (attributes). Besides that, attribute values from one entity may be used by other entities. In these cases, they will assume synthesized or inherited features. From the generic AST, it is possible to define rules to validate the structure that will be applied context-aware system.

Thus, it is possible to state that AGs can contribute to the consolidation of intelligent environments through different approaches. In this paper, it will be addressed the possibility of applying AGs for the validation of activities of users.

3 Grammar for Human Activity Recognition

This section has as the main purpose to present an attribute grammar for validation the structure of the users' activities recognized in an intelligent environment. The grammar has the following structure:

p1: $activity \rightarrow tasks, activityDescription$
p2: $tasks \rightarrow task*$
p3: $task \rightarrow subtasks, taskDescription$
p4: $subtasks \rightarrow subtask*$
p5: $subtask \rightarrow name, location$
p6: $taskDescription \rightarrow name, location, begin$
p7: $activityDescription \rightarrow code, name, location, resource, begin, duration$
p8: $duration \rightarrow TEXT, begin$
p9: $name \rightarrow name1, location$

Considering that every activity performed by users can be seen as a sequence of actions, for this attribute grammar it was defined that an *activity* can be composed of a set of *tasks* and an *activityDescription* (**p1, p2**). These *tasks* refers to small actions that the user has to perform before execute the *activity* itself. Each of these *tasks* may be composed of a set of *subtasks*, which are smaller actions that need to be concluded in preparation for the *task* to be achieved and a *taskDescription* (**p3, p4**). Each *subtask* is composed of its *name* and *location* where it should be performed (**p5**). The *taskDescription* includes the task's *name, location* and *hour* of starting (**p6**). The *activityDescription* refers to the attributes that are considered to be relevant, namely, an identification *code*, the *name* of the activity, the *location* where the user intends to perform it, a set of *resources* that will be necessary for the user to accomplish it, the hour that he started (*begin*) and the *duration* of the activity (**p7**). The duration of an *activity* is defined by the current time (*TEXT*) and the starting time (**p8**). At last, the names of activities result from the names of each *task* that compose them and their *location*. The names of tasks result from the *subtasks*' names and *location* (**p9**).

Figure 2 presents an example applied to a specific activity named *readBook*. It shows an AST that represents how the grammar was developed and how it will be used to validate the activities.

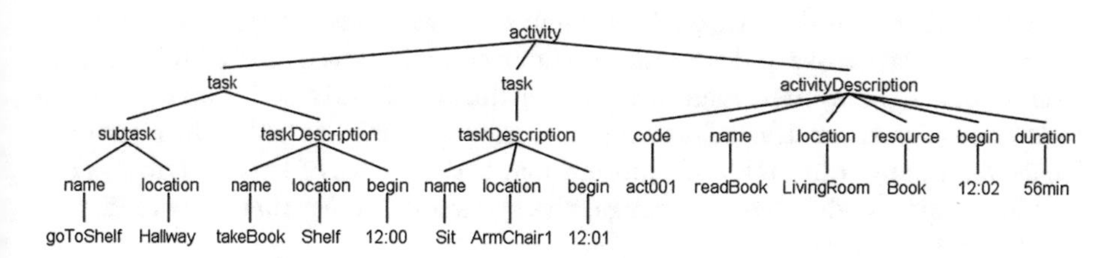

Fig. 2. Abstract Syntax Tree of the human activities recognition grammar.

The values of the attributes in this example refer to data acquired from sensors. After properly analysing this raw data, the system fills the structure that will be used by the grammar to validate the activities. Thus, the attributes of the symbols in the grammar are represented by context data.

These preparation steps that the user accomplish before performing an activity are naturally necessary, thus, it is important for the system to understand them. In Fig. 2, it is possible to identify that the structure of the activity *readBook* has one set with two tasks that must be complete before (and during) the performance and one subtask that represents the preparation for its development.

Considering the structure of the grammar and the values of attributes detected by sensors for the activity of *readBook*, the productions will have the following behaviour:

p5: $subtask \rightarrow goToShelf, Hallway$
p6: $taskDescription \rightarrow takeBook, Shelf, 12:00$
p6: $taskDescription \rightarrow Sit, ArmChair1, 12:01$
p7: $activityDescription \rightarrow act001, readBook, LivingRoom, Book, 12:02, 56\,\text{min}$

All the productions related to terminal symbols in the grammar will have there structure filled by sensor data. As can be seen, **p6** is executed twice, since the activity is composed by two tasks (*takeBook* and *Sit*).

The composition of activities in intelligent environments is performed through a definition of a set of attributes that are considered to be relevant for them to be accomplished. For this, the system should have a module with machine learning algorithms aiming identification of patterns of user's behaviour. Through the understanding of the steps that the user usually performs by accomplishing a set of tasks (activity), it is possible to identify what context data the system should consider to model new activities. For instance, by analysing the user's routine activities the system identifies that usually when he goes to the shelf in the hallway, he also takes a book, sits on the armchair and spends some time there, it is possible to infer that this flow of actions has a considerable probability of being part of an activity. From the analysis of this set of parameters, it is possible to classify it as being part of the activity of reading. The AG will use this data as synthesized and inherited attributes for the symbols for the validation of the structure of the activity.

The validation is done through the analysis of each of the steps that composed them, i.e., if their tasks and subtasks are finished within the expected parameters. Besides that, the system identifies the conclusion of activity by analysing the behaviour of the user. It considers the time used to finish the tasks, the resources, location and any other relevant information. If the values of these attributes are within a predefined acceptable range, the system can infer that the activity as concluded with success.

The mapping and monitoring of activities are important because any changes identified during the execution of an activity, task or subtask, may be related to problems with the user, for example, if considering an Ambient Assisted Living domain, the unexpected changing of behaviour can characterize health problems.

3.1 Semantic Rules

One of the main advantages of applying attribute grammars to intelligent environment relays in the possibility of creating semantic rules that can be used to properly manipulate context data. For the development of this research it was defined that each *activity* is the result of a set of *tasks* and other additional parameters. And, each of the *tasks* is the result of a set of *subtasks* and additional parameters. The following list presents some examples of semantic rules defined for the activities recognition grammar.

1. activityDescription.duration = getCurrentTime(TEXT.value) - activityDescription.begin;
2. activityDescription.begin = getTime(TEXT.value);
3. activityDescription.resource = TEXT.value;
4. activityDescription.location = TEXT.value;
5. activityDescription.name = taskDescription1.name(taskDescription2.name)*,
   taskDescription1.location(taskDescription2.location)*;
6. activityDescription.code = TEXT.value;
7. taskDescription.begin = getTime(TEXT.value);
8. taskDescription.location = TEXT.value;
9. taskDescription.name = subtask1.name(subtask2.name)*,
   subtask1.location(subtask2.location)*;

10. subtask.location = TEXT.value;
11. subtask.name = TEXT.value;

The first rule (1) refers to the calculation of the time spent by the user to finish a specific activity. This rule allows the system to identify changes in the behaviour of the user while performing that activity. Thus, with the information of this parameter, the system could infer the impact that this can have to the user. Besides the *TEXT.value* referring to the current time, this rule uses an inherited attribute related to the starting time of the activity (rule 2). Applying these rules to the activity of *readBook*, the structure is the following:

1. activityDescription.duration = getCurrentTime(TEXT.value) - 12:02;
2. activityDescription.begin = 12:02;

The third rule (3) contains the data related to any resource used by the user to perform the activity. The resources used and his location are fundamental for the system to adapt its applications in order to create a better environment, aiming to assist him. In the *readBook* activity, this rule has the following structure:

3. activityDescription.resource = Book;

The location of the user is relevant for the system to understand his intentions (rules 4, 8 and 10). The name of the activity is defined through the analysis of the tasks involved in preparation to it and where they were performed (rule 5). In the *readBook* activity:

4. activityDescription.location = LivingRoom;
8. taskDescription.location = Shelf;
8. taskDescription.location = Shelf;
8. taskDescription.location = Armchair1;
10. subtask.location = Hallway;

Rules number 2 and 7 allow the grammar to identify the exact time the user started an activity or task. Rules 6 and 11 are generated by the system and in the *readBook* activity will have the structure:

6. activityDescription.code = act001;
11. subtask.name = goToShelf;

The productions of the semantic rules can be used as the basis for the analysis of the level of certainty of specific contexts. This means that they can contribute to the minimization of the problem of uncertainty. For instance, the system should be aware of the time spent by the user to finish the activities. The following code describes how production could be allied with context analysis.

```
if (getDuration(activity.duration) not in expecDurValues[])
then analyse(activity.duration)
```

The result produced by rule 1 should be analysed according to a pre-defined range of values, referring to the time. If the production does not have a value within the expected set, this might mean that there are problems related to the activity. For instance, the value of *sensorData.TIME.now* was not captured with precision, resulting in an uncertain time spent or it may be related to problems with the user, once he is not been able to finish the task as usual. To solve this problem all the process of the *activity.duration* rule should be executed again.

4 Related Work

One of the main contributions of this work is the description of an attribute grammar applied to intelligent environment. At least to the best of the authors' knowledge, there are no similar approaches relating these two fields of research. Thus, aiming to describe the feasibility of using this approach, it is presented in this paper a case study of modelling of activities. There are works in the literature with relevant contributions that can be related to the one presented here, even though not using attribute grammars to model human activities. Most of them apply decomposition of tasks into subtasks and actions, creating a hierarchy and considering temporal aspects.

In [6], the authors states that formal modelling of tasks help users to perform activities with computational support. For instance, it is possible to define rankings of priority of activities to be performed according to the user's context, considering, for example, safety and social skills. The authors consider three parameters: simulation of time and conditions, referring to preconditions for tasks and possible obstacles; spatial behaviour, that analyses the location where the tasks will be performed, the objects used and the possibility of changing actors from one location to another, and; analysing communication, referring to the hierarchy and dependence.

In [2], it is described the syntax and semantics of a function model capable of integrating task analysis with formal verification called EOFM (Enhanced Operation Function Model), based on Extensible Markup Language (XML). Changes related to activities are standardized by the model and it defines the behaviour of them according to their goals, based on the analysis of state and derived variables.

In [8], it is presented the ConcurTaskTree Environment (CTTE), which's goal is to provide means for management of task models for cooperative applications. Among the features, it is interesting to highlight: clear definition of temporal logics for task and their relations; the creation of categories such as user tasks (human routine activities), application tasks (related to the system), interaction tasks (relating user and the system) and abstract tasks (tasks belonging to more than one category) [8].

5 Discussion and Final Considerations

Different formalisms are been used to validate proposals for improvement of intelligent environments. An attribute grammar is an alternative due to the strict formalism for the definition of its symbols and attribute values. It allows the specification of semantic rules providing meaningful characteristics for the grammar. This creates a strong bond with context-aware applications. Thus, it is possible to create another layer of validation for the formal representations of real-world situations.

Another contribution of this work refers to the formalism provided by attribute grammars that can be extended to other ends, for instance, Ambient Assisted living, House Automation and Intelligent Vehicles. This can be evidenced once, in spite of the domain that an attribute grammar is being applied, its structure will have all necessary arguments, e.g., symbols, attribute values and production of semantic rules.

Context-aware systems often face the problem of uncertainty of information. This may be caused by problems with hardware of sensors preventing them to send updated data to be manipulated. Another source is related to design problems. This makes the system to not understand precisely what it should do with context data. Uncertainty can be faced by increasing the amount of context data that is used by the system. For instance, installing more sensors to monitor the environment and its users, or applying new algorithms of machine learning and inference rules aiming to increase the knowledge of the system [10]. Besides that, the use of propositional logic to create rules is a useful approach. However, in many cases it is impossible to map all the consequences for a given premise due to the very wide range of possibilities it can have [9]. According to [3], one way of minimizing uncertainty is through the analysis of knowledge models seeking for all possible alternatives of assumptions, regarding specific contexts and defining probability weights for each of the possible outcomes.

According to [11], the definition of uncertainty models help the identification of its source and the definition of optimal solutions for decision making. They can generate unique outputs based on the processing of imprecise data. This can be applied to context-aware situations formalized through an attribute grammar approach when the data from sensors are not enough to define a situation.

It is believed that attribute grammars can be allied to other fields of artificial intelligence, e.g., machine learning algorithms. Thus, the next step of the project is to merge them aiming the improvement of solutions for context-aware systems. At last, real-world tests are still needed for the validation of this approach.

Acknowledgements. This work has been supported by FCT – Fundação para a Ciência e Tecnologia within the Project Scope: UID/CEC/00319/2019.

References

1. Anderson, M.P., Woessner, W.W., Hunt, R.J.: Forecasting and uncertainty analysis. In: Anderson, M.P., Woessner, W.W., Hunt, R.J. (eds.) Applied Groundwater Modeling, 2nd edn., Chap. 10, pp. 443–491. Academic Press, San Diego (2015). https://doi.org/10.1016/B978-0-08-091638-5.00010-9
2. Bolton, M.L., Siminiceanu, R.I., Bass, E.J.: A systematic approach to model checking human-automation interaction using task analytic models. IEEE Trans. Syst. Man. Cybern. Part A Syst. Hum. **41**(5), 961–976 (2011)

3. Burgman, M.: Risks and Decisions for Conservation and Environmental Management. Ecology, Biodiversity and Conservation. Cambridge University Press (2005). https://doi.org/10.1017/CBO9780511614279
4. Bürger, C., Karol, S., Wende, C.: Applying attribute grammars for metamodel semantics. In: ECOOP 2010 Workshop Proceedings - International Workshop on Formalization of Modeling Languages, FML 2010 (2010). https://doi.org/10.1145/1943397.1943398
5. Freitas, L., Henriques, P.R., Novais, P.: Uncertainty in context-aware systems: a case study for intelligent environments. In: Rocha, Á., Adeli, H., Reis, L.P., Costanzo, S. (eds.) Trends and Advances in Information Systems and Technologies, WorldCist 2018. Advances in Intelligent Systems and Computing, 1 edn., vol. 745, pp. 225–231. Springer International Publishing (2018). https://doi.org/10.1007/978-3-319-77703-0_22
6. Giese, M., Mistrzyk, T., Pfau, A., Szwillus, G., von Detten, M.: AMBOSS: a task modeling approach for safety-critical systems. In: Forbrig, P., Paternò, F. (eds.) Engineering Interactive Systems. Springer, Heidelberg (2008)
7. Knuth, D.E.: Semantics of context-free languages. Math. Syst. Theory **2**(2), 127–145 (1968)
8. Mori, G., Paternò, F., Santoro, C.: CTTE: support for developing and analyzing task models for interactive system design. IEEE Trans. Softw. Eng. **28**(8), 797–813 (2002). https://doi.org/10.1109/TSE.2002.1027801
9. Russell, S., Norvig, P.: Artificial Intelligence: A Modern Approach, 3rd edn. Prentice Hall Press, Upper Saddle River (2009)
10. Tian, W., Heo, Y., de Wilde, P., Li, Z., Yan, D., Park, C.S., Feng, X., Augenbroe, G.: A review of uncertainty analysis in building energy assessment. Renew. Sustain. Energy Rev. **93**, 285–301 (2018). https://doi.org/10.1016/j.rser.2018.05.029
11. White, J.T., Fienen, M.N., Barlow, P.M., Welter, D.E.: A tool for efficient, model-independent management optimization under uncertainty. Environ. Model. Softw. **100**, 213–221 (2018). https://doi.org/10.1016/j.envsoft.2017.11.019

Capturing Play Activities of Young Children to Detect Autism Red Flags

Mariasole Bondioli[1(✉)], Stefano Chessa[1], Antonio Narzisi[2], Susanna Pelagatti[1], and Dario Piotrowicz[1]

[1] Department of Computer Science, University of Pisa, L. Pontecorvo 3, 56127 Pisa, Italy
{mariasole.bondioli,susanna.pelagatti}@di.unipi.it, stefano.chessa@unipi.it, dariopiot@gmail.com

[2] IRCCS Stella Maris Foundation, Calambrone, Italy
anarzisi@inpe.unipi.it

Abstract. The use of IoT devices to monitor activities of users is an established methodology in e-health and ambient assisted living, even if its adoption is still limited to a few, albeit popular, applications. We propose its adoption also in a niche application, namely the observation of young children during their games, which is a common test performed by specialists to diagnose autistic spectrum disorders in an early stage. Specifically, we describe an IoT system that employs miniaturized sensors and data fusion algorithms based on machine learning to identify automatically the movements applied to the toys by the children, and we propose a protocol for its use in the forthcoming pilot experiments.

Keywords: Autism · IoT · Early diagnosis · Neural network

1 Introduction

Observing young children while they play is a test often adopted by specialists to identify in an early stage potential autistic spectrum disorders (ASD) so to trigger specific diagnostic tests [2,11]. The observation of the play is motivated by the fact that ASD children exhibit a behavior in the games usually different from non ASD children. For example, they often carry toys around without playing, and/or they insist in games with very rigid rules, and they often lack social play and imagination in fiction games. At the state of the art, the observation of the play of children requires specialist knowledge and it is conducted in clinics. This, however, is not the ideal setting for the observation of an ASD children since the toys are not theirs and the environment is unusual.

Several recent studies in internet of things (IoT) concerning the recognition of human activities by means of wearable devices [5] suggest that "intelligent" devices embedded into toys and connected to the internet may be trained to remotely provide information about the play of the children in their own environments (at home or at school). Such use of IoT technologies would enable to

P. Novais et al. (Eds.): ISAmI 2019, AISC 1006, pp. 71–79, 2020.
https://doi.org/10.1007/978-3-030-24097-4_9

new protocols in the early identification of red flags concerning ASD, since they may be used without the presence of a specialist (who would have access to this information from the clinic), and they would allow to cover a wider range of ages and a larger number of children. Although the research on activity recognition with IoT devices is already advanced and many commercial products are widely available, most of the existent work are focused on applicative fields concerning fitness or ambient assisted living in general [1].

In our previous work [9] we already addressed the problem of designing an IoT system specific for the detection of movements of toys for children. In that work we focused on the design of a miniaturized sensor with accelerometers, gyroscope and magnetometer to be embedded into toys for young children, and on a basic data fusion algorithm aimed at producing informations about yaw, roll and pitch of the toy. In this work we build over our previous work to develop an improved version of the system.

The novel contribution of the present work comprise an improved data fusion algorithm. Specifically, we introduced an additional layer to the data fusion algorithm that operates in pipeline with the previous one. The new layer, which is based on machine learning, aims at providing a classification of some typical movements (like move forward or backward, etc.) applied by the children to the toy during a play session. This layer implements a neural network operating over both raw accelerometer data and data obtained by the previous layer. The training and the performance evaluation are conducted over a training set and a test set obtained by collecting a dataset simulating short play sessions. The performance evaluation shows the effectiveness of the approach chosen. The paper also present the detailed plans for the pilot studies, that will be strategic to collect a real-world dataset aimed at both improving the classification capability of the system, both at addressing more complex movements for the classification. The pilots will be carried out in cooperation with the hospital IRCCS Stella Maris in Calambrone (Italy).

In the next section we briefly discuss the related works, then, in Sect. 3, we present the new system with its improved data fusion algorithm and interfaces, and in Sect. 4 we show the experimental results validating the data fusion algorithm. Section 5 presents the plan for the experimental pilots and Sect. 6 draws the conclusions.

2 Related Work

The connection between the movements of the children interacting with the daily life objects and the signals of a potential early diagnosis of autism has been investigated and confirmed by several previous research projects.

In recent years, the medical interest in the study of motor impairments in the ASD people has been renewed. At the same time, the growing interest in the ICT field for using technology to ease several aspects of diagnosis and treatment of ASD people has led to a rapid increase of the experimental work based on innovative supports to clinical and educational path totally fitted on the Autistic condition. Thus, we motivate our project of a smart non-invasive diagnosis approach starting from a multidisciplinary background, based on studies both in the medical and ICT research field.

Looking further into this general picture, on the clinical side, the outcome of recent research, conducted also with the help of family video recording, has made it possible to highlight how the compromises on the motor system in autism are frequently observable already at a time of very early development in children who will subsequently receive a diagnosis of ASD [13,14].

In this field, we can even find some studies supporting a clinical approach focused on the direct observation and evaluation of the children's movement. For instance, with their research [3] prove how in young ASD children (mean age 33 months) the lack of social engagement and attention could be traced to their tendency to the prevalent restricted object use measured during the experimentation. In another work, with the observation and interpretation of atypical visual exploratory behaviors toward inanimate objects (AVEBIOs) Mottron et al. highlight the predisposition to look to the objects with lateral glances, as a specific more common in children with ASD [10].

In the multidisciplinary perspective of our work, from the technological point of view we can identify a limited number of previous research exploiting the IoT to monitor and collect informations about the ASD children's activities, no ones specifically focused on early diagnosis purposes. In [6] Goodwin et al. use a sensors system to investigate the automatic detection of stereotypical children movements. Aiming to test a classifier algorithm trained to recognized repetitive behaviors, they obtain a good result in their experimental session, monitoring by video recording and collecting the data for the algorithm analysis by six children wearing three wireless accelerometer, at school and during therapy.

Plotz et al. in their work combine a device with a 3D accelerometer and a microcontroller with machine learning techniques, to demonstrate the accuracy of the machine learning approach to the detection and classification of children anomalous behavior [12].

In other studies, the researchers investigate the recognition of stereotypical movements by the use of the Kinect, as [7] that developed a software to automatically recognize using the machine learning approach the movements filmed by the kinect camera and describe the result of its test on 12 actors performances of three differents stereotypical movements [7].

Differently from the described work, our research aims to test directly on field an IoT system developed for non-expert users. As we will show in the follow sections, in fact, in the first planned experimentation of our system, we will provide to stress the system in a real-life context, in a non-invasive approach.

Moreover, from this first overview emerges that the neural networks exploitation for ASD diagnostic purpose and for the labelling of the children's movements is almost an unexplored but promising field.

3 The Motion Capture System

The motion capture system comprises three main high-level components:

1. **The sensorized toys and the data fusion algorithms** that were developed in our previous work, presented in [9]. The device, implemented with a small-size sensor, is embedded in toys for children to enable a natural children play session.
2. **The Back-end implementation and the data analysis tools** that implements a remote storage in the cloud for the data acquired by all devices in use and a neural networks activity recognition module with an higher-level classification of the movements performed by the children with the sensorized toys.
3. **The User interface** a framework designed to facilitate the remote control of the devices by the clinical staff (to control the measurements sessions), by the system developers (to manage the devices and the training of the neural networks) and provide to the specialists in a intuitive form the results of the data analysis.

Focusing on the sensorized toy, we developed the motion capture device to measure the direction of the movement and the acceleration applied to the toy by a child, implicitly giving also information about if, when and how long the child plays with the toy. The device, developed over the Arduino-based particle photon platform, uses a three axis accelerometer, a magnetometer and a gyroscope and it is powered with a rechargeable powerbank of 2600 mAh. The device performs cycles of data sampling, processing and transmission at 22 Hz, which has been determined empirically as the maximum speed that the device can sustain at an almost constant rate. The data is processed in the device by means of a data fusion algorithm that produces information about the instantaneous roll, yaw and pitch of the toy, and about the force applied. Each data produced by the device is incapsulated into a JSON record and sent by means of the WiFi interface of the device through a router to a remote server by means of the MQTT protocol. We equipped two toys: a truck and an airplane "superwings" with two of such devices to perform our experiments. The devices are placed so to align the X-axis of the accelerometer lenghtwise, the Y-axis in a transverse direction, and the Z-axis vertically.

As anticipated before, the core of the innovation in the new prototype is the introduction of an activity recognition component based on a neural network model implemented with Python and TensorFlow [4]. The neural network structure consists of four layers: one layer for the normalized input values, two hidden layer each with a number of nodes proportional to the input and an output layer.

The devices' recorded data is split in one-second windows each with its set of 22 data samples (from the 22 Hz sampling), which in turn are comprised of 15 numerical values (6 provided by the data fusion algorithm and 9 recorded directly from the on-board sensors). These 330 numerical values (after being refined with a min-max normalization function) constitute the networks' input. The network structure has being defined empirically and evaluated to obtain the best possible trade-off between accuracy and efficiency.

After a training phase and an accuracy evaluation the resulting neural network models, described into a hdf5 output file, are deployed by using the

TensorFlow.js library. To the purpose of training of the neural networks, we collected and labelled a dataset containing sequences of typical movements by using the same prototypal toys, configured to this specific purpose. This dataset includes 6000 sequences labeled manually. For each movement a randomized subset of the dataset was extracted with a 1:2 ratio of positive and negative examples of it (85% of the dataset was used for training and the remaining 15% was used for testing purposes). The training has been performed by using the Adam algorithm (an efficient stochastic optimization algorithm that in the last few years has gained popularity in the Machine Learning community) [8]. The system has currently been tested on 6 different movements (*forward*, *backward*, *up*, *down*, *shake* and *still*) but it has anyway been designed with the capacity of categorizing, if given sufficient examples, any user-provided one-second type of movement.

The remote control of the session and the data collected on the back-end of the system, on the client side are managed by a framework conjugates the final user needs with the developer experimental project intentions. In this prospective, we designed a multifunctional software for the management of the interactions between the users and the system by three different type of actions (see Fig. 1): (1) Patients and session management, (2) Data analysis and output visualizations, (3) Neural network training. The graphical user interface enables the control of the toys by the therapists or the parents, the control the interactions of the children with the toys and the visualization of the data acquired, by providing functions to configurate and initiate new sessions, to add or remove the toys to/from the pool of available toys etc. The dedicated sections allow non-expert users to easily and remotely manage the sensors activities and, at the same time, to orderly collect the informations about the session development.

Secondly, the system returns the data collected and analyzed by means of four different visualization tools: (1) an interface that permits to the user to upload a videotaping of a recorded play session, and remotely synchronize it with the data collected by the sensors; (2) a 3D visualization of the movements applied to the sensorized object flanked by the numerical values (by using the Three.js library); (3) the movement diagrams of a diagnostic session; (4) the neural networks output, classifying the movements of a session as coherent or incoherent.

The application provides means for easy creation and management of the necessary neural networks' support data. Through it the user can create the classification target movements, give them labels and descriptions and produce manually their examples which are needed for the models' training. For the latter the user chooses for which movement it wants to create one or more examples, subsequently specifies the manner of acquisition (either for a single instance or a sequence) and then performs the motion with the specified toy when prompted by the application. Finally, a separate tool allows the developer to train the neural networks based on the aforementioned data, verify their accuracy and generate the hdf5 files for deployment.

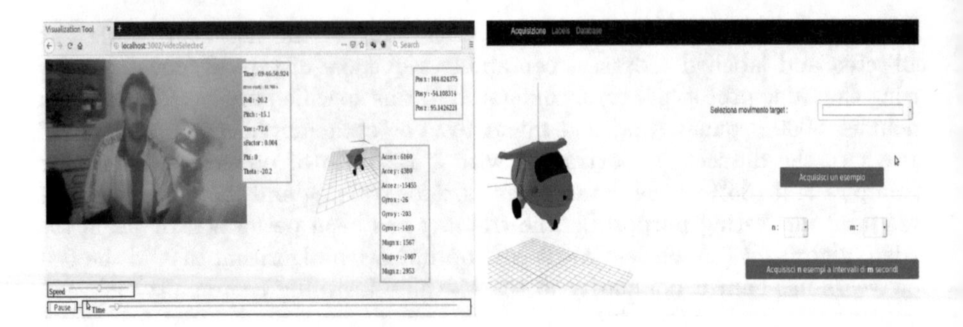

Fig. 1. The sections of data visualization (left) and movements labelling (right)

4 Experimental Results

The neural network models (one binary classifier for each movement) underwent an initial and preliminary phase of testings with the aim of evaluating their accuracy and usefulness. To the purpose of testing, we used 15% of the set of 6000 labeled sequences acquired in the initial data collection campaign. Table 1 shows the various (binary) confusion matrices obtained by the tests the models, where the lines correspond to the label of the sequences (the ground truth) and the columns report the result of the binary classification of the model. In general, classifications have yielded overall positive accuracy results ranging from 0.90 to 0.99 but mixed results for precision, recall and F-measure as shown in Table 2. Not surprisingly, the cases of the classes *Still* and *Shake* are those that give the best results, contrarily the classes *Forward* and *Backward* are the ones deserving more attention given the fact that these two movements appear almost symmetrical but the trained models provide rather different results. In our future work we will devote more experiments to this case.

Table 1. Confusion matrices

Shake

	False	**True**
False	3255	12
True	4	157

Forward

	False	**True**
False	3016	296
True	14	102

Backward

	False	**True**
False	3207	75
True	25	121

Up

	False	**True**
False	2996	277
True	26	129

Down

	False	**True**
False	3040	259
True	24	105

Still

	False	**True**
False	3174	114
True	0	140

Table 2. Accuracy, precision, recall and F-measure values

Movement	Accuracy	Precision	Recall	F-Measure
Shake	0.995	0.928	0.975	0.951
Forward	0.909	0.256	0.879	0.396
Backward	0.970	0.617	0.828	0.707
Up	0.911	0.317	0.832	0.459
Down	0.917	0.288	0.813	0.425
Still	0.966	0.551	1	0.710

5 Planning of Pilot Test

The pilot test of the Motion capture 2.0 prototype is starting during this current month (February 2019), with the already consolidated multidisciplinary collaboration between the computer science university Department of Pisa and the Stella Maris foundation, a public academic hospital with a high level of intervention on the diagnosis and treatment of subjects with Autism Spectrum Disorder.

After over a month of inspections and observations of medical activities in their real context, we have planned together with the medical staff the play session protocol.

The main aims of the pilot test 2.0 will be, not only to create a first on field dataset to give to the trained neural network, but also to validate and verify the system in different real life contexts and to collect the first results about the comparison between the control group movements output and those of the ASD children sample. In this perspective the experimentation will be set in three different location (the ASD diagnosis room of Stella Maris foundation, the private therapy room of one of the therapists involved in the research and a kindergarten classroom), with three different sample groups of children. Indeed, the experimentation foresees, on one side, a sample group of 6 children, from 8 to 10 years old with certified high functioning ASD, already included in a therapy path, with comparable scores in two different cognitive scales (the ADOS-2 and the CBCL), that will use the toys during their regular therapy session. On the other side the system will be tested by a control group of 20 pre-scholar neurotypical children, during the kindergarten time. Finally, we will provide for a third testing group, including randomized children, from 5 to 10 years old, unknown by the therapists, that will use the toys at the end of their first diagnosis visit in the Stella Maris Foundation, to verify if it's possible to cluster the data collected based on the comparison of the preliminaries results of the psychological scale tests and the motion capture output.

Each 10-min session, in each different set, will be controlled remotely by the researcher and the therapist using the software described, and, at the same time, it will be recorded by a camera, thus obtaining the video to be loaded, when requested, by the data analysis tool.

6 Conclusions

The remote data observation and analysis of children movements in natural play contexts permits to face the ASD diagnosis with an innovative approach. With the combination of ubiquitous IoT systems and the automation of the sensor data analysis, the system described in this paper enables an earlier and accurate analysis of children play sessions in their own environment and for long periods, that can be used in clinical and in-home contexts. Although still in a pre-experimental phase, the system positively answered to a first laboratory validation with interesting results during the preliminary tests, that concern the ability to classify correctly some typical movements applied by children to their toys during their games. In this perspective, we already planned the next experimental phase, with a sample of already diagnosed ASD children and a control group of neurotypical children set in two different location: a private medical clinic and a school classroom, to extend our dataset and to validate in a real-life context our multidisciplinary approach.

References

1. Amoretti, M., Chessa, S., Furfari, F., Lenzi, S., Wienapper, F.: Sensor data fusion for activity monitoring in ambient assisted living environments. In: Hailes, S., Sicari, S., Roussos, G. (eds.) Sensor Systems and Software. S-CUBE 2009. LNICST, pp. 206–221. Springer, Heidelberg (2010)
2. Baranek, G.T., Barnett, C.R., Adams, E.M., Wolcott, N.A., Watson, L.R., Crais, E.R.: Object play in infants with autism: methodological issues in retrospective video analysis. Am. J. Occup. Ther. **59**(1), 20–30 (2005)
3. Bruckner, C.T., Yoder, P.: Restricted object use in young children with autism: definition and construct validity. Autism **11**(2), 161–172 (2007)
4. Goldsborough, P.: A tour of TensorFlow. Proseminar Data Mining (2016)
5. Gomez, C., Chessa, S., Fleury, A., Roussos, G., Preuveneers, D.: Internet of Things for enabling smart environments: a technology-centric perspective. J. Ambient Intell. Smart Environ. **11**(1), 23–43 (2019)
6. Goodwin, M.S., Intille, S.S., Albinali, F., Velicer, W.F.: Automated detection of stereotypical motor movements. Autism Dev. Disord. **41**(6), 770–782 (2011)
7. Kang, J.Y., Kim, R., Kim, H., Kang, Y., Hahn, S., Fu, Z., Khalid, M., Schenck, E., Thesen, T.: Automated tracking and quantification of autistic behavioral symptoms using microsoft kinect. Stud. Health Technol. Inform. **220**, 167–170 (2016)
8. Kingma, P., Ba, J.L.: Adam a method for stochastic optimization (2015)
9. Lanini, M., Bondioli, M., Narzisi, A., Pelagatti, S., Chessa, S.: Sensorized toys to identify the early 'red flags' of autistic spectrum disorders in preschoolers, pp. 190–198 (2018)
10. Mottron, L., Mineau, S., Martel, G., Bernier, C.S., Berthiaume, C., Dawson, M., Lemay, M., Palardy, S., Charman, T., Faubert, J.: Lateral glances toward moving stimuli among young children with autism: early regulation of locally oriented perception? Dev. Psychopathol. **19**(1), 23–36 (2007)
11. Ozonoff, S., Macari, S., Young, G.S., Goldring, S., Thompson, M., Rogers, S.J.: Atypical object exploration at 12 months of age is associated with autism in a prospective sample. Autism **12**(5), 457–472 (2008)

12. Ploetz, T., Hammerla, N.Y., Albinali, F., Rozga, A., Abowd, G.: Automatic assessment of problem behavior in individuals with developmental disabilities, pp. 391–400 (2012)
13. Provost, B., Heimerl, S., Lopez, B.R.: Levels of gross and fine motor development in young children with autism spectrum disorder. Phys. Occup. Ther. Pediatr. **27**(3), 21–36 (2007)
14. Radonovich, K.J., Fournier, K.A., Hass, C.J.: Relationship between postural control and restricted, repetitive behaviors in autism spectrum disorders. Front. Integr. Neurosci. **7**, 28 (2013)

Social Robots with a Theory of Mind (ToM): Are We Threatened When They Can Read Our Emotions?

Jin Kang(✉) and S. Shyam Sundar(✉)

Media Effects Research Laboratory,
Donald P. Bellisario College of Communications,
The Pennsylvania State University, University Park, PA 16802, USA
{jbk5361, sss12}@psu.edu

Abstract. How would human users react to social robots that possess a theory of mind (ToM)? Would robots that can infer their users' cognitions and emotions threaten their sense of uniqueness and evoke other negative reactions because ToM is a uniquely human trait? If so, can we alleviate these negative user reactions by framing robots as members of our ingroup? We addressed these questions with a 3 (robot's affective ToM: correct vs. incorrect vs. control) × 2 (robot's group membership: ingroup vs. outgroup) × 2 (user gender: female vs. male) between-subjects online experiment. Participants were asked to complete an online task with a robot named Pepper that was identified as an ingroup member or outgroup member. They first read a passage describing a past user's interaction with Pepper, in which the user expressed sarcasm and Pepper correctly or incorrectly identified the user's sarcasm or made a neutral comment. Males reacted more negatively to Pepper that correctly identified sarcasm and reported lower expected enjoyment with Pepper than females. Ingroup Pepper made participants feel closer to the robot but also threatened their sense of uniqueness than the outgroup Pepper. Design implications for fostering better human-robot interaction (HRI) are discussed.

Keywords: Theory of mind · Threat to uniqueness · Social robots · Ingroup

1 Introduction and Related Work

Theory of mind (ToM) consists of two components, cognitive and affective [1]. The former is the ability to understand and predict others' behaviors by inferring their thoughts, knowledge and intentions (i.e., one's belief about others' beliefs), while the latter is the ability to understand and predict others' behaviors by inferring their emotional states (i.e., one's beliefs about others' emotions) [1]. Generally, ToM is considered a uniquely human trait because of the depth and speed with which humans are able to demonstrate and adopt this trait compared to nonhuman species [2]. Endowing social robots with affective ToM has been a great interest among robotics researchers. It is believed that such robots can foster truly empathetic interactions between human users and robots because they can personalize their subsequent action to address the functional and emotional needs of each user [3, 4]. Social robots that

P. Novais et al. (Eds.): ISAmI 2019, AISC 1006, pp. 80–88, 2020.
https://doi.org/10.1007/978-3-030-24097-4_10

detect a user's frustration at a given task can adjust the task's difficulty to reduce their level of frustration. Similarly, social robots that can pick up sarcasm from a user can dynamically change the interaction to ensure a better user experience. But, would users be comfortable with robots that can read their mind and emotions?

Some studies suggest that the answer is "no" because such behaviors may be seen as eerily uncanny. The "uncanny valley effect" in human-robot interaction refers to the phenomenon in which a robot's human likeness positively influences users' emotional responses up to a certain point after which its human likeness provokes aversion [5]. Several researchers argue that this repulsion towards overly humanlike robots is due to the fact that these robots undermine human's uniqueness [6–9]. Humans are unique in that we look and think markedly differently from other species. Seeing social robots that possess some of the defining human features blurs the line between humans and machines. As a result, these robots threaten users' unique identity as human and evoke a host of other negative reactions (e.g., eeriness) [6–10].

We therefore predicted that human users would not like social robots that demonstrate affective ToM (i.e., robots that correctly identify users' true emotions). Such robots would threaten users' sense of unique identity and evoke negative reactions. Instead, we surmised that human users would prefer robots that fail to use affective ToM (i.e., incorrectly identify users' true emotions), because they would serve to reassure them of their unique human identity. Hence, we proposed the following hypothesis:

Hypothesis 1 (H1). Human users will (a) feel greater threat to uniqueness, (b) experience lower psychological closeness and (c) expect lower interaction enjoyment after seeing social robots with accurate affective ToM compared to seeing robots with inaccurate affective ToM.

If our prediction is correct, what would be one solution to mitigate users' resistance toward social robots with affective ToM? We propose that users' negative responses can be mitigated by framing social robots as belonging to their ingroup. Individuals like ingroup members more than outgroup members because the ingroup provides them with meaningful social identity [11]. Humans show positive bias towards their ingroup members, including recalling more positive information about the members [12] and enacting more cooperative and prosocial behaviors [12, 13]. Such positive ingroup bias has also been observed in HRI. Individuals who belonged to the same group as social robots felt closer towards the robots [14], expressed greater willingness to interact with the robots, and perceived greater warmth in the robots [15]. Given this, we hypothesized that social robots' group membership can effectively shape human users' reactions towards the robots with affective ToM. With ingroup social robots, users will feel less threat to their unique identity and report other positive reactions after seeing the robots correctly using affective ToM, because it would be acceptable—perhaps even expected—for ingroup members to share their unique traits. On the contrary, for outgroup social robots, the opposite effects are hypothesized. That is, users would report feeling threatened and experience other negative reactions when they see the robots correctly using affective ToM as opposed to incorrectly using affective ToM. Formally stated:

Hypothesis 2 (H2). There will be a two-way interaction between a robot's affective ToM and a robot's group membership in influencing human users' (a) sense of threat to uniqueness, (b) psychological closeness and (c) expected interaction enjoyment.

Additionally, we investigated the role of the user's gender in influencing their reactions towards social robots with affective ToM. Gender is an important user characteristic that determines users' attitudes and interests towards social robots [16, 17]. Thus, we posed a research question probing for the unique interaction between user's gender, robot's affective ToM and robot's group membership:

Research Question (RQ). What is the relationship between robot's group membership, user's gender and user responses towards social robots with affective ToM?

2 Method

We conducted a 3 (robot's affective ToM: correct vs. incorrect vs. control) × 2 (robot's group membership: ingroup vs. outgroup) × 2 (user's gender: female vs. male) between-subjects online study. In the study, participants were told they would work on an online task called Desert Survival Task [18] in the same or different team as Pepper. Before the task, they were asked to read a passage that depicted how Pepper typically interacted with human users. This passage described an unpleasant interaction between two human users where one user expressed sarcasm. Pepper correctly or incorrectly identified the user's sarcasm or made a neutral comment. Decoding sarcasm requires one to exercise their affective ToM [19], and our use of sarcasm was designed to demonstrate the robot's capacity for affective ToM. We included Pepper that made a neutral comment (i.e., control) to assess users' baseline reaction towards the robot without any affective ToM. We used vignettes to induce indirect interactions with social robots, which has been shown to be effective in robotics research [6, 9].

2.1 Participants

A total of 220 participants were recruited from Amazon Mechanical Turk for a compensation of $ 0.75 (Male = 45.7%; $Mean_{age}$ (M) = 35.23, $Standard\ Deviation_{age}$ (SD) = 10.61). MTurk workers are more representative of the adult population compared to other sample bodies (e.g., students) [20], thus increasing the external validity of our findings. The distribution of race in the sample was as follows: White (78.9%), African American (8.6%), Asian (5.9%), Other (5.4%) and American Indian (1.1%). After deleting those who failed the manipulation check for robot's group membership, our final dataset had 185 participants.

2.2 Independent Variables

Affective ToM was operationalized as whether Pepper correctly or incorrectly identified the story character's sarcasm. In the control condition, Pepper made a neutral comment. *Robot's Group Membership* was operationalized as whether Pepper belonged to the participant's team for an upcoming online task. In the ingroup condition, participants were told Pepper belonged to their team, and they should cooperate with Pepper to complete the task. In the outgroup condition, participants were told Pepper did not belong to their team, and they would compete against Pepper for the task.

2.3 Procedure

After obtaining informed consent, participants were told that they would be assigned to the same or different team as Pepper to complete a Desert Survival Task [18]. This task was said to be an online task where they would interact with Pepper via online video chat service. To enhance the credibility of the cover story, participants were told they needed a computer with a camera to see Pepper online. Then, participants saw a short video of Pepper (15 s), introducing itself. Next, participants read a passage that described how Pepper typically interacted with humans in the real world before meeting Pepper online. The passage described that Pepper had been deployed to a home of two graduate students named Sam and Jenn for a week. The passage depicted a scene in which Sam and Jenn were trying to decide which cupcake to eat – chocolate or vanilla. Sam loved chocolate cupcakes, and Jenn knew about Sam's preference. However, as Sam was reaching for a chocolate cupcake, Jenn snatched the cupcake from her hands. Sam made a frustrated facial expression and commented, "Wow! You are the best roommate ever! Thank you for being so considerate! Isn't it great, Pepper?". In the correct affective ToM condition, Pepper replied, "It seems like you are not very happy right now. Can I play your favorite song to make you feel better?" successfully detecting Sam's sarcasm. In the incorrect condition, Pepper replied, "It seems like you are very happy right now. Can I play your favorite song while you eat the cupcake?" failing to detect Sam's sarcasm. In the control condition, Pepper replied, "Yes? Do you need my assistance? What can I do?" Then, participants were directed to the survey platform Qualtrics to complete the outcome measures.

2.4 Dependent Variables

Unless otherwise indicated, we used a 7-point Likert scale (1 = strongly disagree; 7 = strongly agree) for all measures. *Manipulation check* for participants' perception of robot's affective ToM was assessed using 3 items, "Pepper can accurately perceive how one truly feels," "Pepper can correctly identify what one is emotionally going through," and "Pepper can detect sarcasm" (Cronbach's α = 93). For robot's group membership, participants were simply asked, "Are you in the same team as Pepper?" with Yes/No response options. *Threat to Uniqueness* was assessed at individual and group levels. Sample items at the individual level are "I feel less unique," and "I cannot think of many special characteristics that make me special" (Cronbach's α = .92). Sample items at group level are "Human's uniqueness is somewhat threatened by robots like Pepper" and "Robots like Pepper undermine what it means to be a human" (Cronbach's α = .91) [9]. *Perceived Closeness* towards a robot was assessed using 3 items, "I am similar to Pepper," "I identify with Pepper," and "I feel connected to Pepper" (Cronbach's α = .95). Participants' *Expected interaction enjoyment* with Pepper for the upcoming task was assessed using 5 items, such as "I will enjoy interacting with Pepper" (Cronbach's α = .81).

3 Result

Unless otherwise indicated, we conducted a 3 (robot's affective ToM: correct vs. incorrect vs. control) × 2 (robot's group membership: ingroup vs. outgroup) × 2 (user's gender: female vs. male) ANOVA on all dependent-variable measures.

3.1 Manipulation Check

Using one-way ANOVA, we found a significant main effect of robot's affective ToM on participants' perception of robot's affective ToM, F (2, 182) = 69.26, $p < .001$, partial $\eta^2 = .43$. As expected, participants in the correct condition agreed more that Pepper can identify one's true emotions (*Mean* (*M*) = 5.07, *Standard Error* (*SE*) = .19) than participants in the incorrect (*M* = 2.29, *SE* = .18) and control (*M* = 2.51, *SE* = .18) conditions. The latter two conditions did not differ from each other.

3.2 Threat to Uniqueness

H1 predicted that robot with accurate affective ToM would lead to higher levels of threat to uniqueness, H2 predicted that robot's group membership would be an important moderator, and RQ asked if user gender had any impact. A 3 (robot's affective ToM: correct vs. incorrect vs. control) × 2 (robot's group membership: ingroup vs. outgroup) × 2 (user gender: female vs. male) MANOVA was conducted with both threat measures as dependent variables. There was a significant main effect of robot's group membership, Wilks'Λ = .96, F (2, 171) = 3.95, $p < .05$, partial $\eta^2 = .04$. Close inspection of the univariate analysis revealed that participants in the ingroup condition felt greater threat to individual uniqueness (*M* = 2.25, *SE* = .13) than participants in the outgroup condition (*M* = 1.79, *SE* = .13), F (1, 172) = 6.22, $p < .05$, partial $\eta^2 = .04$. Providing response to RQ and partial support to H1, there was a significant interaction between robot's affective ToM and user gender, Wilks'Λ = .94, F (4, 342) = 2.64, $p < .05$, partial $\eta^2 = .03$. A univariate analysis, F (2, 172) = 4.42, $p < .02$, partial $\eta^2 = .05$, revealed that male participants in the correct (*M* = 2.44. *SE* = .22) and incorrect conditions (*M* = 2.36, *SE* = .25) felt greater threat to individual uniqueness than those in the control condition (*M* = 1.61, *SE* = .23). For female participants, however, there were no significant differences among the three conditions (correct condition: *M* = 1.65, *SE* = .23; control: *M* = 2.11, *SE* = .21; incorrect conditions: *M* = 1.95, *SE* = .21).

There was a marginally significant interaction between robot's affective ToM and user gender for threat to human uniqueness, F (2, 172) = 2. 43, $p = .09$, partial $\eta^2 = .03$. Male participants in the correct condition (*M* = 2.81. *SE* = .23) felt greater threat compared to those in the incorrect (*M* = 2.48, *SE* = .25) and control (*M* = 1.92, *SE* = .24) conditions. Female participants did not differ across conditions (correct condition: *M* = 2.07, *SE* = .24; control condition: *M* = 2.16, *SE* = .21; incorrect condition (*M* = 2.45, *SE* = .21) (Fig. 1). H2 was not supported for both threat measures as evidenced by the non-significant two-way interaction between robot's affective ToM and robot's group membership, F (2, 172) = .70, $p = .50$, partial $\eta^2 = .01$ (threat to

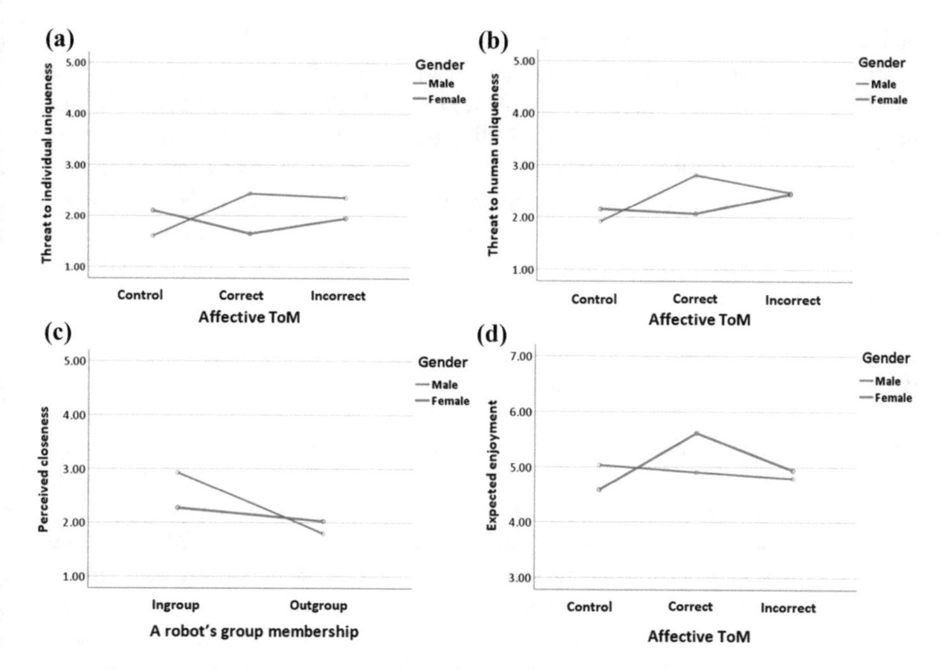

Fig. 1. Summary of significant interaction effects: (a) threat to individual uniqueness, (b) threat to human uniqueness, (c) perceived closeness, and (d) expected enjoyment.

individual uniqueness) and F (2, 172) = 1.16, p = .32, partial η^2 = .01 (threat to human uniqueness).

3.3 Perceived Closeness and Expected Enjoyment

H1 predicted that a robot with accurate affective ToM would lead to lower levels of closeness and expected enjoyment, while H2 predicted that the robot's group membership would be an important moderator. There was a significant main effect of robot's affective ToM, F (2, 172) = 8.28, p < .001, partial η^2 = .09. Those in the correct condition felt closer to Pepper (M = 2.79, SE = .16) than those in the control (M = 2.08, SE = .16) and incorrect conditions (M = 1.89, SE = .17). The main effect of robot's affective ToM was not significant for expected enjoyment, F (2, 172) = 1.93, p > .05, partial η^2 = .02. H1 was not supported. There was also a significant main effect of robot's group membership, F (1, 172) = 13.40, p < .001, partial η^2 = .07. Those in the ingroup condition felt closer to Pepper (M = 2.60, SE = .14) than those in the outgroup condition (M = 1.91, SE = .13). This main effect is qualified by the interaction between user gender and robot's group membership, F (1, 172) = 5.54, p < .05, partial η^2 = .03. Males in the ingroup condition reported feeling closer to Pepper (M = 2.93, SE = .20) than those in the outgroup condition (M = 1.80, SE = .19). For females, the effect of robot's group membership on perceived closeness was not strong (ingroup: M = 2.27, SE = .19; outgroup: M = 2.02, SE = .17). In response to RQ, there was a marginally significant interaction between robot's affective ToM and user's gender for expected enjoyment, F (2,172) = 2.75, p = .07, partial η^2 = .03. While males' expected enjoyment was not significantly influenced by robot's affective ToM (control: M = 5.04, SE = .26; correct: M = 4.91, SE = .25; incorrect conditions: M = 4.79, SE = .27), females' expected enjoyment was higher for the correct condition (M = 5.61, SE = .26) than the incorrect (M = 4.94, SE = .23) and control conditions (M = 4.58, SE = .23). H2 was not supported (two-way interaction between robot's affective ToM and robot's group membership for perceived closeness, F (2, 172) = 1.79, p = .17, partial η^2 = .02; expected enjoyment, F (2, 172) = 1.108, p = .33, partial η^2 = .01).

4 Discussion and Conclusion

Our data showed that participants' gender was an important factor shaping their responses towards robot with affective ToM. Male participants found Pepper that correctly and incorrectly used ToM threatening to their individual uniqueness. On the contrary, female participants felt that the Pepper that correctly used ToM as the least threatening and even expressed the greatest expected interaction enjoyment with this Pepper. One explanation for this gender difference is that our participants are from the U.S. where gender roles are quite prominent. Males in western societies are socialized to be competent in various tasks while females are socialized to be socially oriented. Perhaps, our male participants saw Pepper that correctly used ToM as taking away their uniqueness of being a competent male and also saw Pepper that incorrectly used ToM as a potential rival. For female participants, having Pepper that can correctly use affective ToM is probably seen as more beneficial as it can facilitate the formation of meaningful interactions. We also found that robot's group membership did not mitigate threat to uniqueness elicited by robot that showed affective ToM –correctly and incorrectly– and that being in the same team (vs. different team) as Pepper evoked higher levels of threat to individual uniqueness. Our significant main effect of robot's group membership on perceived closeness gives us confidence that our manipulation of this independent variable worked successfully, especially for male participants. These results suggest that our participants felt mixed emotions being in the same team as Pepper: their uniqueness was threatened, yet they felt closer to the robot. Perhaps, the coexistence of such mixed emotions towards ingroup robots means the robot's group membership is not an effective method to reduce the threat associated with robots that show affective ToM.

Based on our results, we recommend robotics researchers to prioritize examining factors that can reduce threat to users' sense of uniqueness as they interact with robots with affective ToM. Considerable design effort is directed at developing systems to enable the robots to recognize users' basic emotions through multiple cues, such as facial expression, and body gestures [3]. Unintentionally, however, robots that show affective ToM may detract from effective user experience, especially for male users, as they can elicit negative user reactions, including feeling a threat to their uniqueness. In order to reduce such reactions, designers would benefit by personalizing robot interactions based on user characteristics.

Against the growing trend to imbue social robots with human traits, our study provides valuable contribution in understanding whether social robots should become more humanlike. We invite future researchers to explore which humanlike features are acceptable by users and which ones are threatening, so that we can successfully design social robots that can have help users in their daily lives and foster harmonious human-robot interactions, partnerships and relationships.

References

1. Shamay-Tsoory, S.G., Aharon-Peretz, J.: Dissociable prefrontal networks for cognitive and affective theory of mind: a lesion study. Neuropsychologia **45**(13), 3054–3067 (2007)
2. Call, J., Tomasello, M.: Does the chimpanzee have a theory of mind? 30 years later. Trends Cogn. Sci. **12**(5), 187–192 (2008)
3. Hong, A.: Human-Robot Interactions for Single Robots and Multi-Robot Teams. MA Dissertation. University of Toronto, CA (2016)
4. Stephan, A.: Empathy for artificial agents. Int. J. Soc. Robot. **7**(1), 111–116 (2015)
5. Mori, M.: The uncanny valley. Energy **7**, 33–35 (1970)
6. Gray, K., Wegner, D.M.: Feeling robots and human zombies: Mind perception and the uncanny valley. Cognition **125**(1), 125–130 (2012)
7. MacDoman, K.F., Vasudevan, S.K., Ho, C.C.: Does Japan really have robot mania? Comparing attitudes by implicit and explicit measures. AI Soc. **23**(4), 485–510 (2009)
8. Kaplan, F.: Who is afraid of the humanoid? Investigating cultural differences in the acceptance of robots. Int. J. Hum. Robot. **1**(3), 465–480 (2004)
9. Ferrari, F., Paladino, M.P., Jetten, J.: Blurring human–machine distinctions: anthropomorphic appearance in social robots as a threat to human distinctiveness. Int. J. Soc. Robot. **8**(2), 287–302 (2016)
10. Stein, J.P., Liebold, B., Ohler, P.: Stay back, clever thing! Linking situational control and human uniqueness concerns to the aversion against autonomous technology. Comput. Hum. Behav. **95**, 73–82 (2019)
11. Hogg, M.A.: Social identity theory. In: McKeown, S., Haji, R., Ferguson, N. (eds.) Understanding Peace and Conflict Through Social Identity Theory, pp. 3–17. Springer, Switzerland (2016). https://doi.org/10.1007/978-3-319-29869-6
12. Howard, J.W., Rothbart, M.: Social categorization and memory for in-group and out-group behavior. J. Pers. Soc. Psychol. **38**(2), 301–310 (1980)
13. Levine, M., Prosser, A., Evans, D., Reicher, S.: Identity and emergency intervention: how social group membership and inclusiveness of group boundaries shape helping behavior. Pers. Soc. Psychol. Bull. **31**(4), 443–453 (2005)
14. Häring, M., Kuchenbrandt, D., André, E.: Would you like to play with me?: How robots' group membership and task features influence human-robot interaction. In: Proceedings of the 2014 International Conference on Human-Robot Interaction, pp. 9–16. ACM, Germany (2014)
15. Eyssel, F., Kuchenbrandt, D.: Social categorization of social robots: anthropomorphism as a function of robot group membership. Br. J. Soc. Psychol. **51**(4), 724–731 (2012)
16. Heerink, M.: Exploring the influence of age, gender, education and computer experience on robot acceptance by older adults. In: Proceedings of the 6th International Conference on Human-Robot Interaction (HRI'11), pp. 147–148. ACM, New York (2011)
17. Arras, K.O., Cerqui, D.: Do we want to share our lives and bodies with robots? A 2000 people survey: a 2000-people survey. Technical Report, pp. 1–41 (2005)

18. Nass, C., Fogg, B.J., Moon, Y.: Can computers be teammates? Int. J. Hum. Comput. Stud. **45**(6), 669–678 (1996)
19. Shamay-Tsoory, S.G., Tomer, R., Aharon-Peretz, J.: The neuroanatomical basis of understanding sarcasm and its relationship to social cognition. Neuropsychology **19**(3), 288 (2005)
20. Buhrmester, M., Kwang, T., Gosling, S.D.: Amazon's mechanical turk: a new source of inexpensive, yet high-quality, data? Persp. Psychol. Sci. **6**(1), 3–5 (2011)

Blockchain-Based Architecture: A MAS Proposal for Efficient Agri-Food Supply Chains

Yeray Mezquita[1(✉)], Alfonso González-Briones[1(✉)], Roberto Casado-Vara[1(✉)], Pablo Chamoso[1(✉)], Javier Prieto[1(✉)], and Juan Manuel Corchado[1,2,3,4(✉)]

[1] BISITE Research Group, University of Salamanca, Calle Espejo 2, 37007 Salamanca, Spain
{yeraymm,alfonsogb,rober,chamoso,javierp,corchado}@usal.es
[2] Air Institute, IoT Digital Innovation Hub (Spain), 37188 Carbajosa de la Sagrada, Spain
[3] Department of Electronics, Information and Communication, Faculty of Engineering, Osaka Institute of Technology, Osaka 535-8585, Japan
[4] Pusat Komputeran dan Informatik, Universiti Malaysia Kelantan, Karung Berkunci 36, Pengkaan Chepa, 16100 Kota Bharu, Kelantan, Malaysia

Abstract. Logistics services involve a wide range of transport operations between distributors and clients. Currently, the large number of intermediaries are a challenge for this sector, as it makes all the processes more complicated. This paper proposes a system that uses smart contracts and blockchain technology to remove intermediaries and speed up logistics activities. Our model combines smart contracts and a multi-agent system to improve the current logistics system by increasing organization, security, transparency and significantly improving distribution times.

Keywords: Blockchain Technology · Multi-Agent System · Smart contract · Supply chain · Agri-food commodities · Logistical utilities

1 Introduction

Logistics services between farmers and retailers in the agri-food supply chain involve a wide range of transport and storage operations. These operations are performed by a large number of intermediaries who create a huge discrepancy between prices that farmers get and the retailers' prices.

The gap produced in the agri-food supply chain process between farmers and final consumers is due to the ever-increasing input cost that stakeholders along the supply chain add to the agri-food commodities [1].

In the Food Price Monitoring Tool [2] is shown that, for example in the case of Spain with a magnitude of 38.5% while assuming that an average agri-food

P. Novais et al. (Eds.): ISAmI 2019, AISC 1006, pp. 89–96, 2020.
https://doi.org/10.1007/978-3-030-24097-4_11

commodity receives an increase of the 20% in its price, the producer would increment just a 7.7% the commodity price. The same happens in other continents like in Asia [3].

To create a transparent and more efficient agri-food supply chain, this work proposes a distributed blockchain-based architecture. Blockchain Technology (BT) allows to keep a tamper-proof record of agri-food commodities ensuring the authenticity of the data recorded. Thanks to this record it is possible to trace back the origin of a commodity and its state during the transactions within the supply chain [4,22].

Within the scope of the agri-food financial market, the use of BT and smart contracts can reduce the cost of transaction while improving the cash flow and the working capital for farmers and sellers [16]. This is possible because the automatic and real-time payments performed through smart contracts eliminate the need for financial intermediaries and the complex processes that normally slow down the flow of assets within the supply chain [8].

The benefits of an optimized supply chain would have a direct impact on the profits of the producers. This is because two thirds of the final cost of a commodity come from the operating cost of the supply chain [9]. Thanks to that, producers could spend more on improving their infrastructures and optimizing their production method, something that is not always possible in the traditional agri-food supply chains [10].

The most promising method of improving agri-food traceability, distributing subsidies or improving transparency is the application of BT to those processes [5–7]. Also, the record of transactions generates a vast amount of data within the supply chain, these data could also provide more accurate market information [5].

To make use of the possibilities that BT brings with itself in agri-food supply chains, this work proposes an architecture in which a Multi-Agent System (MAS) helps coordinate the actors of a blockchain-based agri-food supply chain [17,20].

This paper is structured as follows: the next section reviews state-of-the-art architectures that combine BT and supply chains. Then, we describe the architecture approach whose aim is to improve and optimize the agri-food supply chains combining BT and the MAS paradigm. Finally, this work concludes with a discussion about the possibilities of the implementation of this approach.

2 Related Work

Crops and agri-food commodities have to be tokenized in order to become part of the transactions registry of a blockchain. The tokenization of a real asset is the creation of an entry in the blockchain with the distinctive details and properties of the real asset, alongside its current owner.

Currently there are not many companies dedicated to tokenizing crops and agri-food products. These companies, in addition to offering their own tokenization protocol, function as regulated markets for the exchange of these products.

AgriDigital is a cloud computing platform built over BT, for the management of digitalized agri-food assets [11]. This platform is focused on creating secure payments for sellers in the supply chain. It has more than 1300 users and more than 1.6 million tons of grain have been traded through the system [12]. They are building a protocol for digital assets that makes use of smart contracts to create an efficient supply chain designed for global commodity trade, finance and traceability.

Fetch is a decentralized platform, which combines a MAS with BT in order to perform an economic work over a digital representation of the real world [13]. The MAS is dynamically reorganized as it continually generates new agents every time they are required. The digitalized world inside its blockchain allows to trace back the digitalized assets. While in combination with the MAS it is possible for the different actors to trade the information they need in an autonomous and unsupervised way.

Binkabi develops blockchain based solutions for agri-food supply chains with the aim of providing value to those actors of the chain who provide worth to a product and cut out those that don't [14]. The actors that don't add real value to a product are the intermediaries' back-to-back trades or those that loan money to a producer and add interests to the value of the product.

To tokenize a commodity in the Binkabi blockchain it is necessary to make use of the protocol of Barter Block platform [15] The user that wants it, has to deposit the grain in an accredited warehouse, it weighs, grades and packs the grain. Warehouse receipts (commodity -backed tokens) are issued to the Depositor.

Sellers post their tokenized commodities in the blockchain while buyers post their offers. The platform matches the orders and make the automatic swap of commodity-backed tokens for fiat-backed tokens. When a holder wants to redeem the commodities, they send the redemption request to the warehouse that stores them. Then the warehouse redeems the commodities and the commodity-backed tokens are burnt.

3 Proposed Architecture

The proposed architecture is based on the MAS paradigm to automatize the organization of a platform. BT is the central element of the model, being used to interconnect and allow the communication between different parts of the platform.

The blockchain is a decentralized network of nodes, while being the centralized part of the proposed system means that our approach doesn't contain a single point of failure, making it a truly decentralized model.

3.1 Blockchain Network

The type of blockchain used in this model is a permissioned one. It is maintained just by a network of nodes belonging to the actors participants of the supply chain. The control of the blockchain network is equally shared between the nodes of the network. To let a new node take part of the network (aspirant) all the nodes have to know its identity and how it takes part in the supply chain process, to make a decision about its admittance and try to reach consensus on it.

If the identity of an aspirant is not known, e.g. it is an attacker that wants to take part in the control of the network, or the aspirant is known, but is an actor that wants to increase its control power of the blockchain, the participants should reach a consensus to reject its admittance in order to maintain the status quo of the blockchain shared control.

Clients that acquire the final product and do not take part in the supply chain process can obtain permissions from the nodes to read data from the blockchain but not to write in it. The agri-food assets data is stored in the blockchain following the non-fungible Ethereum ERC-721 token standard.

Smart contracts written in solidity (https://solidity.readthedocs.io/en/v0.5.5/) that implements this kind of interface let the edition of features of each asset individually. This characteristic allows the status of the digitalized asset to be continuously updated in an automated way through out the different stages of the supply chain.

Like in the case of Binkabi, all the tokens that represents a real state, to be valid, must meet the following requirements:

1. Guaranteed: a token must be guaranteed by a correct quantity of the product that back it up.
2. Auditable: outside the blockchain there have to be audits of the assets that back up the tokens inside the platform. To help automate this process it is possible to make use of the Internet of Things (IoT) devices that automatically insert the data collected and measured from the real world inside the blockchain. The devices that insert data from the outside into the blockchain are called oracles, they only insert the data that are required by a smart contract [23]. Periodic audits guarantee the existence of the assets and the accuracy of the oracles.
3. Legally enforceable: The right of token investors to agri-food assets must be recognized by the law. For example, if the token holder wishes to exchange the token for another type of product, then he can do so in the accredited secondary market without any problems.

3.2 MAS Design and Real World Interaction

The MAS approach of our work is similar to the one proposed on Fetch, being the main difference between those architectures the way the agents negotiate and reach agreements in order to optimize the operation processes of their platform. In our approach, the agents make use of an auction process in order to obtain the most profit from an asset or to spend the minimum amount of resources to obtain it. The agents are developed in the Java Agent DEvelopment Framework (JADE http://jade.tilab.com/).

A network of sensors reads the state of the real world and digitalizes it. Moreover, the Radio Frequency Identification (RFID) tags and reader antennas improve the automation of the platform [19, 21].

As shown in Fig. 1, the architecture consists of four layers in which agents are grouped on the basis of their function inside the system with the blockchain as the central element of the communication between agents in different layers.

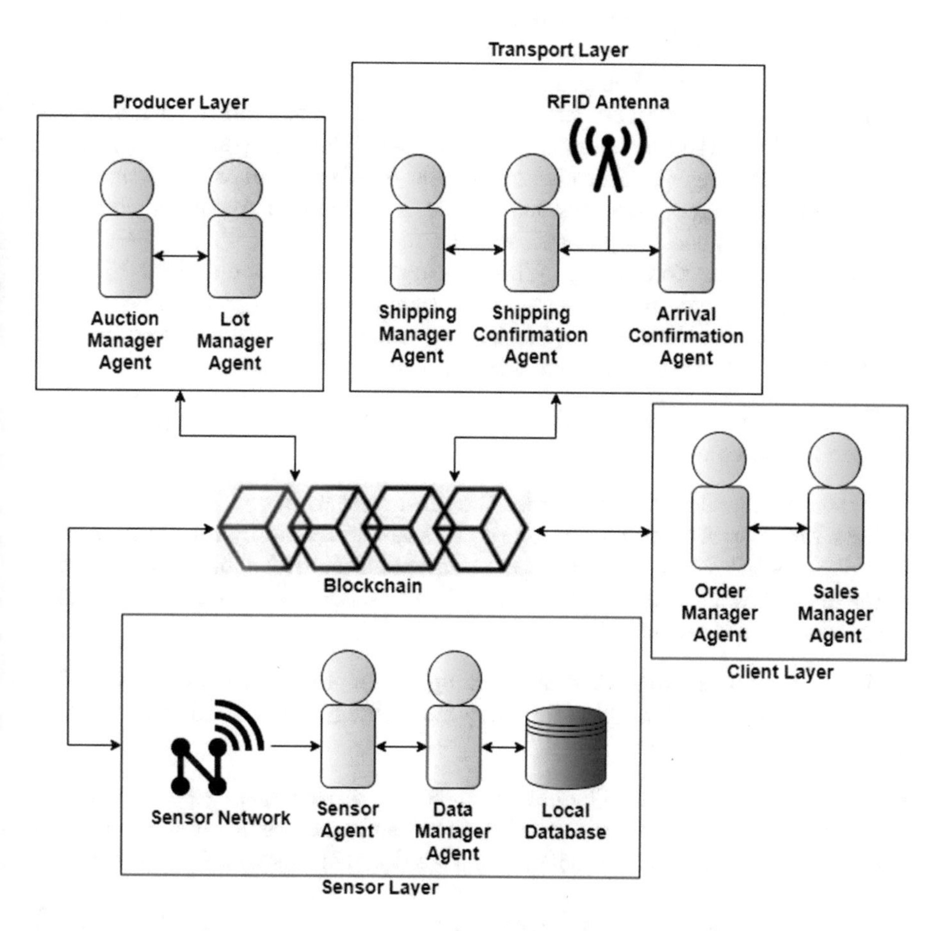

Fig. 1. MAS Architecture of the supply chain platform.

In the Producer Layer, producers tokenize their lots of agri-food commodities and put them on sale:

- The Lot Manager Agent: producers make use of this agent to tokenize the agri-food commodities that are being sold in the platform. This agent makes use of a smart contract that creates tokenized products inside the blockchain. It also tells the Auction Manager which lots are being put on sale.

- Auction Manager Agent: Checks the offers that arrive at the blockchain for each lot on sale. When the time of the auction ends, it makes use of a smart contract, to sell the digitalized lot of agri-food commodities to the highest bidder.

The Transport Layer corresponds to the agents that take care of the movements of the agri-food commodities within the supply chain:

- Shipping Manager Agent: Users interact with this agent to tell the platform that a token is being shipped. This agent tells the Shipping Confirmation Agent which token is being shipped.
- Shipping Confirmation Agent: When an agri-food commodity with an RFID tag, that is marked as being shipped in the blockchain, pass near the RFID Antenna, this agent confirms the details of where and when it has been shipped. This agent also tells the Arrival Confirmation Agent the commodities that have been shipped successfully.
- Arrival Confirmation Agent: When an agri-food commodity that has been successful shipped, passes near the Antenna of a marked destination, this agent marks the token in the blockchain as having arrived successfully at its destination. Everything is done through a smart contract.

The agents that interact with the network of sensors of the supply chain are grouped in the sensor Layer:

- Sensor Agent: Gets the data from the sensor network.
- Data Manager Agent: This agent receives the data read by the sensor Agents and stores them in a local database. If it has an internet connection, it dumps the data stored locally in the blockchain and associates it with the tokenized objects in its location.

The client Layer groups the platform agents in charge of searching for offers in the blockchain, making bids for them and selling them outside the agri-food commodity supply chain:

- Order Manager Agent: The clients make use of this agent to search for the offers in the market of a commodity with the characteristics they need. Then, it makes a bid in the blockchain associated to the tokenized commodity via smart contract.
- Sales Manager Agent: Every asset sold by the client outside the blockchain is managed by this agent. The agent makes use of a smart contract to update the state of the agri-food asset as a sold item that has already left the supply chain.

4 Conclusions

This work proposes a MAS-based architecture that integrates BT within its platform for efficient and transparent agri-food supply chains. The MAS system is in charge of automating and optimizing the interactions between actors that are involved in the supply chain. To do so, agents manipulate the blockchain through smart contracts implemented in solidity, tokenizing and updating the information associated with the agri-food commodities that are registered in the platform following the ERC-721 standard for non-fungible tokens.

The use of BT is justified because it can create transparent and efficient interactions between the actors of the system. Also, it is a decentralized tamper-proof registry that allows for the establishment of trustful agreements between actors. Finally, it is the central actor used by the agents of the MAS to localice and communicate with other parts of the system. The use of a blockchain as a central actor, helps avoiding single point failures during the life cycle of the system.

The platform also makes use of RFID technology and sensors to improve the automation of interactions of the commodities in the supply chain. The use of this technology helps reduce the human error that could appear in the agri-food supply chain while enhancing the flow of assets within it.

The scientific contribution of this paper is the designed platform architecture; unlike other state-of-art platforms its architecture is fully distributed, since it has no central server and therefore no single point of failure. The interactions of the agents are coordinated through the blockchain while they work on the intelligent optimization of the processes carried out within the system.

The implementation of smart contracts for the flow of assets within the platform help in getting rid of the human error while automatize the processes in the supply chain. The smart contracts also take care of the supervision of the conditions that actors must carry out in their agreements, significantly improving distribution times and costs.

Acknowledgments. This work was developed as part of "Virtual-Ledger-Tecnologies DLT/Blockchain y Cripto-IOT sobre organizaciones virtuales de agentes ligeros y su aplicación en la eficiencia en el transporte de última milla", ID SA267P18, project cofinanced by Junta Castilla y León, Consejería de Educación, and FEDER funds. The research of Yeray Mezquita is supported by the pre-doctoral fellowship from the University of Salamanca and Banco Santander.

References

1. Ait Sidhoum, A., Serra, T.: Volatility spillovers in the Spanish food marketing chain: the case of tomato. Agribusiness **32**(1), 45–63 (2016)
2. Food Price Monitoring Tool. https://ec.europa.eu/eurostat/cache/infographs/foodprice/index.html. Accessed 22 Feb 2019
3. Jongwanich, J., Wongcharoen, P., Park, D.: Determinants of Consumer Price Inflation versus Producer Price Inflation in Asia. Asian Development Bank Economics Working Paper Series 491 (2016)
4. Tian, F.: An agri-food supply chain traceability system for China based on RFID & blockchain technology. In: 2016 13th International Conference on Service Systems and Service Management (ICSSSM). IEEE (2016)
5. Tripoli, M., Schmidhuber, J.: Emerging Opportunities for the Application of Blockchain in the Agri-food Industry. FAO and ICTSD: Rome and Geneva. Licence: CC BY-NC-SA 3 (2018)

6. Green, S.: Decentralized Agriculture: Applying Blockchain Technology in Agri-Food Markets. MS thesis. Faculty of Graduate Studies (2018)
7. Galvez, J.F., Mejuto, J.C., Simal-Gandara, J.: Future challenges on the use of blockchain for food traceability analysis. TrAC Trends in Analytical Chemistry (2018)
8. Bacchi, U.: UN Glimpses into Blockchain Future with Eye Scan Payments for Refugees. Thomson Reuters Group 21 (2017)
9. Niforos, M.: Beyond Fintech: Leveraging Blockchain for More Sustainable and Inclusive Supply Chains. EMCompass Note 45. International Finance Corporation (World Bank Group), Washington DC, September 2017. www.ifc.org/wps/wcm/connect/a4f157bb-cf24-490d-a9d4-6f116a22940c/EM+Compass+Note+45+final.pdf?MOD=AJPERES
10. Phillips, L.: How agri-food value chains work, 30 July 2012. https://www.farmersweekly.co.za/agri-business/agribusinesses/how-agri-food-value-chains-work. Accessed 22 Feb 2019
11. AgriDigital. https://www.agridigital.io/blockchain. Accessed 25 Feb 2019
12. Kamilaris, A., Fonts, A., Prenafeta-Boldú, F.X.: The Rise of the Blockchain Technology in Agriculture and Food Supply Chain
13. Simpson, T., Sheikh, H., Hain, T., RNønnow, T., Ward, J.: Fetch: Technical Introduction. https://fetch.ai/uploads/technical-introduction.pdf. Accessed 25 Feb 2019
14. Binkabi. https://www.binkabi.io/. Accessed 25 Feb 2019
15. Barter Block. https://www.binkabi.io/barter-block. Accessed 25 Feb 2019
16. Casado-Vara, R., González-Briones, A., Prieto, J., Corchado, J.M.: Smart contract for monitoring and control of logistics activities: pharmaceutical utilities case study. In: The 13th International Conference on Soft Computing Models in Industrial and Environmental Applications, pp. 509–517. Springer, Cham, June 2018
17. González-Briones, A., De La Prieta, F., Mohamad, M., Omatu, S., Corchado, J.: Multi-agent systems applications in energy optimization problems: a state-of-the-art review. Energies **11**(8), 1928 (2018)
18. González-Briones, A., Valdeolmillos, D., Casado-Vara, R., Chamoso, P., Coria, J.A.G., Herrera-Viedma, E., Corchado, J.M.: GarbMAS: simulation of the application of gamification techniques to increase the amount of recycled waste through a multi-agent system. In: International Symposium on Distributed Computing and Artificial Intelligence, pp. 332–343. Springer, Cham, June 2018
19. Casado-Vara, R., Chamoso, P., De la Prieta, F., Prieto, J., Corchado, J.M.: Non-linear adaptive closed-loop control system for improved efficiency in IoT-blockchain management. Inf. Fusion **49**, 227–239 (2019)
20. González-Briones, A., Castellanos-Garzón, J.A., Mezquita Martín, Y., Prieto, J., Corchado, J.M.: A framework for knowledge discovery from wireless sensor networks in rural environments: a crop irrigation systems case study. Wireless Commun. Mobile Comput. (2018)
21. Casado-Vara, R., de la Prieta, F., Prieto, J., Corchado, J.M.: Blockchain framework for IoT data quality via edge computing. In: Proceedings of the 1st Workshop on Blockchain-enabled Networked Sensor Systems, pp. 19-24. ACM, November 2018
22. Casado-Vara, R., Prieto, J., De la Prieta, F., Corchado, J.M.: How blockchain improves the supply chain: case study alimentary supply chain. Procedia Comput. Sci. **134**, 393–398 (2018)
23. Curran, B.: What are Oracles? Smart Contracts, Chainlink & "The Oracle Problem". Accessed 15 Feb 2019

Sensing as a Service: An Architecture Proposal for Big Data Environments in Smart Cities

Diego Valdeolmillos(✉), Yeray Mezquita(✉), and Alberto R. Ludeiro(✉)

BISITE Research Group, University of Salamanca, Calle Espejo 2, 37007 Salamanca, Spain
{dval,yeraymm,albertoludeiro}@usal.es

Abstract. This paper proposes an architecture capable of responding to the acquisition, processing and storage of information using as reference data of the Smart City, for this purpose it will suggest the use of certain technologies to be used together to meet the needs of a Smart City. A new Cloud Computing paradigm will be used, Sensing as a Service increasing the amount of data recovered and processed to add more value to the system. It proposes the creation of an open, flexible, extensible and self-adaptive architecture in a Big Data environment, capable of providing the acquisition and processing of large volumes of information while maintaining reliability and availability, as well as allowing easy adaptation in terms of scalability.

Keywords: Smart City · Big Data · Sensing as a Service

1 Introduction

Since the dawn of the digital era the growth of information is increasing exponentially, currently we leave a fingerprint on everything we do that involves a digital transaction, more than 5 billion people use their mobile phones to generate information, 100 TeraBytes of data are uploaded to Facebook every day, 571 new websites are created every minute [1]. In addition, the amount of information produced by machines is also growing by leaps and bounds, the IoT where smart devices communicate with each other or with servers. Now we can find smart vehicles, machines in companies and sensors placed by cities collecting and transmitting data, and this will increase due to the lower cost of sensors.

Technologies capable of acquiring, processing, storing and visualizing large volumes of data are needed, and that is where the expression Big Data (BD) comes in, the human being needs to help himself/herself of these technologies in order to be able to manage the information, and to be able to make decisions about business, cities, and in general to help the people, being able to analyze these data through predictive and/or descriptive algorithms [2].

With the rise of BD, SCs and Internet of Things technologies, a new paradigm within the Cloud Computing (CC) called Sensing as a Service (SenaaS) emerges,

P. Novais et al. (Eds.): ISAmI 2019, AISC 1006, pp. 97–104, 2020.
https://doi.org/10.1007/978-3-030-24097-4_12

following the growth of the development and decrease in sensor prices. This is based on whether the owners of the sensors be it a public or private organization or a private person may have a platform for the publication of the data collected by their sensors to obtain some benefit or not [3]. To serve these new needs of Smart City (SC), an open, distributed, scalable and extensible architecture is proposed to enable the acquisition of large volumes of information from heterogeneous sources, its storage and processing. Components of the architecture must be easily integrated, capable of handling large volumes of data in real time, but also using a persistence layer for batch processing [4].

2 Internet of Things

Internet of Things (IoT) is a concept that has been developed with the rise of the Internet in which more and more everyday objects are connected, these objects communicate with human beings as well as with other devices [5]. This is allowing for an increase in the information transmitted about people and their environment, supposedly to make their lives easier in the future. With the reduction in the price of sensors, their number is increasing and they are being installed in work environments, devices of daily life, homes and public and private places, so that any object in the next few years will be a data source.

3 Open Architectures in IoT

When introducing open architectures in IoT framework, these must include components that can be easily extended and integrated, and that are compatible and interchangeable between them, providing flexibility in the choice of components according to the specific functionality to be obtained [6], being significant the use of code and open standards. Sensor generators are the main data source on a IoT platform, but other types of data sources (information from social networks, web page crawler, open data services) can also be included to add value to IoT platforms. With the existence of multiple heterogeneous data sources, ETL (Extraction - Transform - Load) processes responsible for the extraction of data from the sources, their transformation and analysis are essential for obtaining new values [7], isolating and cleaning those problematic values and loading data into the persistence layer [8]. It is necessary to implement open standards for the interpretation of this data that allow interoperability between components, which is defined by the IEEE (Institute of Electrical and Electronic Engineers) as the ability of two or more systems or components to exchange information and use the information exchanged [9]. Therefore, it is inevitable that open representation formats such as XML, JSON, RDF or CSV are used in the transversal nature of the architecture. It is convenient to use some open communication standard for the delivery of messages from sensor generators, the existing IoT platforms include APIs REST over HTTP or HTTPS, MQTT (Message Queue Telemetry Transport), CoAP (Constrained Application Protocol), among others.

4 Sensing as a Service

In recent years, the growth of sensor development has increased dramatically due to advances in sensors, the decrease in their production price and the evolution of CC technology [10]. IoT is based on the fact that a variety of objects can communicate and interact with each other in order to find a common objective, these objects communicate through the Internet or use radio frequency identification (RFID) devices, Bluetooth, WiFi or other communication protocols and can be sensors, mobile phones, actuators. The IoT will allow persons and objects to use existing networks and services to be interconnected at all times and places with any other person or objects.

SenaaS model focuses on the fact that the owner of a sensor, whether a private person, public or private organization, government, etc., can make sensor data available for publication with some benefit [10]. The SenaaS model is based on four conceptual layers [11]:

- Sensor and Sensor Owners layer. A sensor is a device that detects or measures any physical property, records it and/or transmits it properly. The sensors are classified into four categories:
  - Personal or domestic. All objects that do not belong to public or private organizations will be equipped with sensors in the near future.
  - Private organizations or places. Objects belonging to an organization or private company or located in a private place. These objects and those belonging to the previous category could be published or not, maintaining privacy and private security, but providing data that may be useful for this SenaaS model.
  - Organizations or public places. Objects belonging to public places and public infrastructures. This sensor data will be published depending on the government's privacy policies.
  - Commercial sensor data providers. Some companies or individuals may deploy and manage sensors on public or private property, agreeing on legal terms with the permission of private property owners or the competent legal authority on public property. There may be an economic agreement between sensor suppliers and property owners [12].

  The owner of the sensor will be able to decide if it publishes the data collected by it, in that case it will register the sensor in a publisher of sensors, being able to specify conditions, limitations, conditions and profitability of some type on the part of the publishers of sensors, be it an economic agreement or of another type [13].
- Sensor publishers Layer. This layer includes the discovery of available sensors, communication with sensor owners and authorization to publish data obtained by sensors in the cloud. A very important part of this layer is the compensation to the owners of the sensors, and the users who consume this data.
- Extended Service Providers layer. They provide new services for data-consuming users, making it possible for sensor publishers to provide services.

Providers of these services are able to interact with multiple publishers of sensors by acquiring more data, even processing it by providing more information to provide more information to the consumer. This feature will increase the price of the service.
- Sensor Data Consumers layer. Applications that consume the endpoint data offered by sensor publishers or service providers.

Some middleware solutions that adapt to the conceptual layers of the SenaaS model can be OpenIoT, Xively and FIWARE

- OpenIoT (Open Source cloud solution for IoT) [14], is an open source middleware that collects data from sensors and distributes them using an on-demand delivery model.
- Xively [15], provides an enterprise platform to gain business value from connected objects by providing services for web or mobile applications that enable communication with devices, user access and integrate business intelligence systems.
- FIWARE [16]. Open software platform used to easily build applications and services in the cloud. It integrates interfaces and open standards which facilitate the publication of IoT services, also allows the analysis and processing of large amounts of data thus providing an added service [17]. Therefore, it fulfills the role of an extended service provider within the SenaaS model.

5 Proposed Architecture

It is proposed to create an open-source architecture that is capable of providing the acquisition and processing of large volumes of information while maintaining reliability and availability, as well as allowing easy adaptation in terms of scalability. The proposed architecture meets the core cluster requirements for BD processing:

- Horizontal scaling: All components have been tested on the inclusion of several nodes horizontally.
- Robustness: All the tools used natively incorporate mechanisms that implement a resilient behavior in case of failures thanks to the replication of nodes, data, computation processes or the combination of these.

The architecture described in Fig. 1 is divided into 3 layers: physical layer, persistence layer and application layer.

5.1 Physical Layer

This layer displays the components of the distributed cluster, combines the elements for data acquisition, cluster resource manager and distributed processing framework. The elements described below compose this layer:

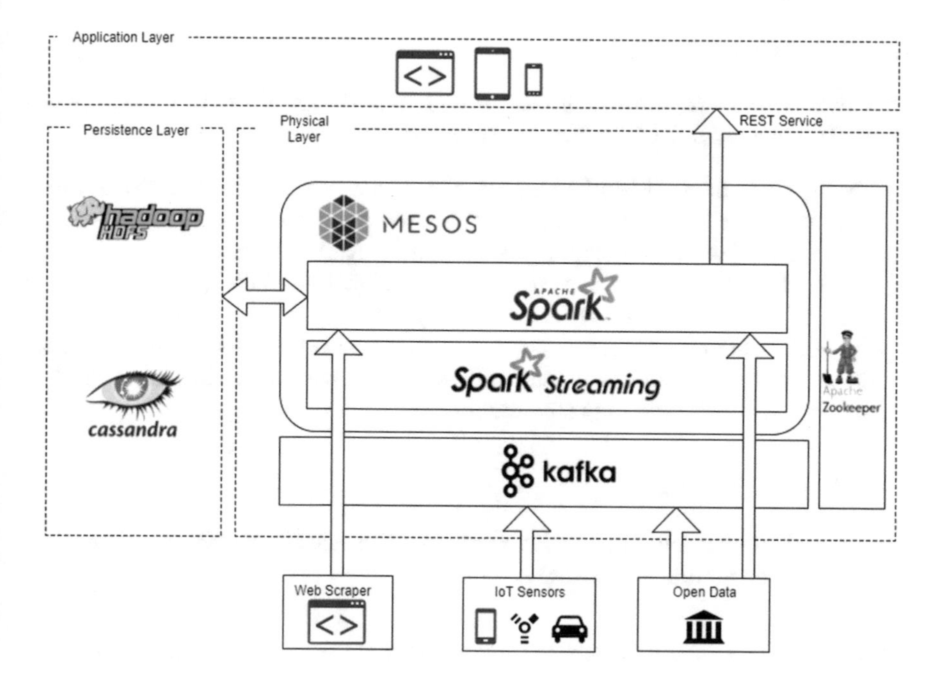

Fig. 1. Proposed architecture

- Apache Kafka. Distributed and partitioned streaming platform based on the publication-subscription model that allows you to publish and subscribe to record streams, store and process them as they happen.
- Apache Spark. Distributed computing framework that uses main memory for its computations, is capable of processing information in real time or batches. It is used for the processing of information collected in a distributed way, processed using the own Machine Learning libraries and the parallelization of readings and scripts in the distributed Hadoop file system (HDFS) and in the NoSQL Cassandra database.
- Spark Streaming. Apache Spark native tool for real-time mass data capture, ensuring high performance, high scalability and fault tolerance. In this particular case Spark Streaming uses Apache Spark to retrieve data via streaming in real time.
- Apache Mesos. A container-based cluster resource manager using master and slave nodes, manages all physical or virtual machine resources connected to the cluster, enabling efficient application deployment. It is used to execute the Spark application on several nodes.
- Apache Zookeeper. Open source distributed service for the coordination of distributed applications, allowing applications to exempt from the responsibilities of synchronization, configuration, management of name groups. Used by Apache Mesos and Apache Kafka.

5.2 Persistence Layer

This layer shows the components of the distributed cluster used for data persistence. The elements described below compose this layer:

- Apache Cassandra. It is a NoSQL database that uses a hybrid model between a column-oriented database and a key-value database. It is used for storing information from the consumption of Apache Kafka streams.
- Hadoop Distributed File System (HDFS). A distributed file system on several nodes of a cluster designed for the storage of large files. It is highly scalable and fault tolerant. It is necessary to use it because Kafka deployed in cluster format needs to temporarily save all records of the topic partitions, storing among other data the sequence number.

5.3 Application Layer

This layer contains any application that uses the REST service, whether it's a web browser or mobile applications.

5.4 Adaptation of SenaaS Model to the Proposed Architecture

The proposed architecture follows the different 4 layers of the SenaaS model [10]:

- Sensors and Sensor Owners layer. Sensors used by citizens and organizations. The sensor information is sent to the Apache Kafka publisher.
- Sensor Publishers layer. Apache Kafka is included for real-time data streaming adoption, making the architecture flexible in the sense of being independent of data formatting. This data obtained in real time will be subsequently exposed through a web service.
- Extended Service Providers layer. The platform allows data capture from heterogeneous sources (Open Data, web crawlers, etc). In addition to the acquisition of data from these sources, inherent in this layer, we add functionality consisting of publishing the information collected via web service (publication layer).
- Sensor Data Consumers layer. The data published by the platform can be consumed, in our specific case they are consumed by the web application.

5.5 Principal Features of the Architecture

The main features that make this architecture robust for the problem are:

- The selected components have already been tested in different scenarios by the community and are widely used.
- It allows a wide variety of information processing scenarios, allowing the implementation of different processing models (batch, real-time, lambda architecture).

- The components that make up the architecture are highly scalable, thus allowing in the case of needing more nodes add them without large extra configurations, being most of the components scalable without the need to turn off the service.
- Cluster mode architecture components are fault-tolerant, making the architecture offer high availability and all implement their own mechanisms to recover and not lose information in the event of a node occurring. For this they use data replication, load balancing and/or message passing semantics.
- In order to capture, process and store data, we have chosen components with high performance in managing large volumes of information.
- Unified resource management under Apache Mesos. Supports Mesos containers, allowing the isolation of tasks.
- It allows the use of different languages in the applications (Scala, Java, Python).
- Mesos, HDFS, and Spark offer web interfaces that allow monitoring the resources used, applications deployed, data stored and information processed, as well as captured through Streaming.
- All components of the architecture offer configurable features to increase security in communications, authentication and user authorization and monitoring.

6 Conclusions

In this article has proposed an architecture capable of resolving communication from sensors adapting to a SenaaS model, which not only allows the publication of sensors but also includes a set of tools for the ingestion, processing, storage and publication of large volumes of data from these sensors.

The components of an open architecture provide flexibility, therefore, must allow you to add and change the components of the architecture avoiding the rigidity of other architectures and whose components must be compatible with each other.

The ecosystem of technologies offered by Apache to respond to different needs around the BD is very diverse and growing day by day, these platforms have characteristics when defining an open distributed architecture because they are open source and flexible being compatible with each other, so they have been taken into account when defining and implementing the architecture.

Acknowledgments. This research has been partially supported by the European Regional Development Fund (FEDER) within the framework of the Interreg program V-A Spain-Portugal 2014–2020 (PocTep) under the IOTEC project grant 0123 IOTEC 3 E. The research of Yeray Mezquita is supported by the pre-doctoral fellowship from the University of Salamanca and Banco Santander.

References

1. Big Data - Interesting Statistics, Facts and Figures (2017). https://www.waterfordtechnologies.com/big-data-interesting-facts/. Accessed 31 July 2017
2. García, O., Chamoso, P., Prieto, J., Rodríguez, S., de la Prieta, F.: A serious game to reduce consumption in smart buildings. In: International Conference on Practical Applications of Agents and Multi-Agent Systems, pp. 481–493. Springer, Cham (2017)
3. Chamoso, P., de la Prieta, F.: Swarm-based smart city platform: a traffic application. ADCAIJ: Adv. Distrib. Comput. Artif. Intell. J. **4**(2), 89–98 (2015)
4. Chamoso, P., González-Briones, A., Rodríguez, S., Corchado, J.M.: Tendencies of technologies and platforms in smart cities: a state-of-the-art review. Wirel. Commun. Mobile Comput. **2018** (2018)
5. Xia, F., Yang, L.T., Wang, L., Vinel, A.: Internet of things. Int. J. Commun. Syst. **25**(9), 1101 (2012)
6. González-Briones, A., Chamoso, P., De La Prieta, F., Demazeau, Y., Corchado, J.: Agreement technologies for energy optimization at home. Sensors **18**(5), 1633 (2018)
7. Casado-Vara, R., Chamoso, P., De la Prieta, F., Prieto, J., Corchado, J.M.: Non-linear adaptive closed-loop control system for improved efficiency in IoT-blockchain management. Inf. Fusion **49**, 227–239 (2019)
8. Grover, R., Carey, M.J.: Data ingestion in AsterixDB. In: EDBT, pp. 605–616 (2015)
9. Geraci, A., Katki, F., McMonegal, L., Meyer, B., Lane, J., Wilson, P., Radatz, J., Yee, M., Porteous, H., Springsteel, F.: IEEE Standard Computer Dictionary: Compilation of IEEE Standard Computer Glossaries. IEEE Press, Piscataway (1991)
10. Zaslavsky, A., Perera, C., Georgakopoulos, D.: Sensing as a service and big data. arXiv preprint arXiv:1301.0159
11. Perera, C., Zaslavsky, A., Christen, P., Georgakopoulos, D.: Sensing as a service model for smart cities supported by internet of things. Trans. Emerg. Telecommun. Technol. **25**(1), 81–93 (2014)
12. Chamoso, P., González-Briones, A., Rivas, A., De La Prieta, F., Corchado, J.M.: Social computing in currency exchange. Knowl. Inf. Syst. 1–21 (2019)
13. Casado-Vara, R., Prieto-Castrillo, F., Corchado, J.M.: A game theory approach for cooperative control to improve data quality and false data detection in WSN. Int. J. Robust Nonlinear Control **28**(16), 5087–5102 (2018)
14. OpenIoT - Open Source cloud solution for the Internet of Things (2016). http://www.openiot.eu/. Accessed 16 Feb 2016
15. IoT Platform for Connected Devices—Xively by LogMeIn (2017). https://www.xively.com/. Accessed 16 Feb 2016
16. FIWARE (2016). https://www.fiware.org/. Accessed 16 Feb 2016
17. González-Briones, A., Chamoso, P., Yoe, H., Corchado, J.: Greenvmas: virtual organization based platform for heating greenhouses using waste energy from power plants. Sensors **18**(3), 861 (2018)

Design of an AI-Based Workflow-Guiding System for Stratified Sampling

G. Hernández(✉), D. García-Retuerta, P. Chamoso, and A. Rivas

Bisite Research Group, Universidad de Salamanca,
Calle Espejo 2, 37007 Salamanca, Spain
{guillehg,dvid,chamoso,rivis}@usal.es

Abstract. The characterization of the resistance of transmission towers is a difficult and costly procedure which can be mitigated using statistical techniques. A stratified sampling process based on the characteristic of the terrain was shown in previous works to reduce the error in the statistical inference; however, such characteristics are usually unknown before a measure is made. In this work, we present a system which integrates artificial intelligence techniques, such as k-nearest neighbors, decision trees, or random forests, to automatically optimize the workflow of expert workers using various sources of data.

Keywords: Ambient artificial intelligence · Statistical sampling · Transmission towers

1 Introduction

The maintenance of high-voltage electric power transmission systems is a topic that has generated a wide variety of research lines [1–3]. In the vast majority of the developed countries, the transmission towers (TTs) that transport the high-voltage lines by air must be periodically checked, according to their characteristics. There are different parameters that must be measured in these revisions, such as, for example, the resistance of the TT, the resistivity of the terrain, the step voltage ($K_{\rm p}$) and the touch voltage ($K_{\rm c}$).

These revisions represent a significant cost since they require highly qualified personnel to work in remote places of difficult access, carrying expensive and heavy equipment. However, part of the values measured in the revision could be predicted. This is because most of the TTs share the same characteristics and are located in similar areas, so in many cases, in addition, they will show similar behavior.

Therefore, the possibility of reducing the cost associated with this type of maintenance is not only attractive, but also quite reasonable. Thanks to the continuous advances of technology, different approaches have been applied to the maintenance processes in different areas or industries, due to the common interest of reducing costs, successfully achieving the expected results. For this reason, the present case study is proposed.

There are four common types of maintenance for TTs [4]:

P. Novais et al. (Eds.): ISAmI 2019, AISC 1006, pp. 105–111, 2020.
https://doi.org/10.1007/978-3-030-24097-4_13

- Reactive maintenance, consisting of solving faults or existing problems to make the system work properly again. Reactive maintenance can be unplanned if problems are fixed at random, without the need of previous intervention or planning, or planned if it is corrected intentionally because it was detected by predictive or preventive maintenance.
- Preventive maintenance, which tries to reduce the amount of equipment failing by looking for solutions to the different problems before they occur. This makes it possible to avoid unplanned reactive maintenance. Although this type of maintenance is focused on increasing reliability and reducing costs, it does not guarantee that failures do not occur in the future. Preventive maintenance has been previously applied in different works such as in [5] to deal with the maintenance of the power lines.
- Predictive maintenance, which emerges as a complement to reactive and preventive maintenance. In this type of maintenance, a series of parameters is monitored and analyzed to determine possible anomalies. The process depends on the ability to generate estimates or assumptions about the state of a given component. When well-defined processes are predicted, especially in the context of control theory [6], it is possible to generate a mathematical model that provides a reliable representation of the real system [7]; however, in other types of processes, machine learning techniques are sometimes required, such as the use of classification algorithms [8] or Artificial Neural Networks [9]. In general, these approaches attempt to extract and model the system using historical data.
- Proactive maintenance, which is a strategy used to maintain the stability and performance of a machine [10], extending the useful life of the equipment while avoiding errors and faults. There are two types of repairs [11]: perfect maintenance, when a machine returns to its ideal state (in this case the cost is usually high), and imperfect maintenance, where an acceptable component of the quality in the repair is sacrificed to reduce the cost. Support Vector Machines (SVMs) are used in this type of maintenance, for example in [12], where they are applied to predict the amount of ice that will accumulate in the high voltage lines. This is a serious problem that can interrupt the electric service for a significant time and its solution can be quite expensive. Another point to take into account in this type of maintenance is the performance of the machine, which, due to its natural deterioration over time, should receive periodic reviews throughout its useful life. To this purpose, periodic maintenance can be performed, periodically evaluating the performance of the machine even though its status is correct.

  This solution is not optimal when the period of the revisions is short and the machines work in perfect conditions. An alternative is based on monitoring of the state of the machinery and the evaluation of different parameters. A combination of these two options is presented in [13], where the loss of performance is predicted from the failure rate and performance degradation.

  Other works, such as [14] analyze failures in the current transformers by applying Dissolved Gas Analysis (DGA). In [15] principal component analysis (PCA) and backpropagation artificial neural networks (BP-ANN) were applied in combination.

The current work focuses on predictive maintenance, where different techniques are already being applied: some authors have used ANN to model the environment, such as in [9], while others use ANN to determine failure times of the devices [16].

In this work, a Virtual Organization capable of carrying out the predictive maintenance of the TTs is proposed for the proactive maintenance of a sample of TTs, which should be reviewed by the corresponding company when a new review period begins. At that time, the operators must indicate to the system which line or lines they should review and that will be the input of the system. The system will execute the different algorithms and functionalities to determine a sample with the TTs that should be reviewed of all the existing TTs in that line.

The algorithms executed by the system for the selection of the TTs to be reviewed are completely autonomous and based on different parameters of the TTs, making it possible to determine the status of all the selected TTs by means of statistical inference. Furthermore, the system is able to predict different values that are measured in the review process, making it possible to determine the best location or the best model when including a new TT.

For this purpose, statistical sampling techniques combined with machine learning methods have been used to estimate the soil resistivity, in addition to the K_p and K_c values. In addition, the system provides different geo-positioning tools to facilitate the search and selection of TTs and lines that are to be reviewed. The system proposed is described in more detail below.

2 Model and Methods

The general purpose of the automated system is to select the TTs to measure, optimizing the capabilities of the statistical techniques eventually applied to subsets of resistance measures. This can be formally defined as follows.

The resistance of a TT can be regarded as a uniformly continuous random variable with a probability density function (pdf) f, such that

$$\int_0^R f(r)\,\mathrm{d}r \tag{1}$$

represents the fraction of TT with a Resistance lower than R. Stating that a resistance will never surpass a certain value—like, for example, the resistance of the human body—can be stated by means of a statistical test, like a binomial one if non-parametric methods are used [17], or more specific tests if a certain family of random variables is assumed to be describing the random variable [18].

Instead of directly performing the inference on this random variable, which exhibits a wide range of variation, according to the data obtained in previous works [19,20], stratified sampling can be done using other parameters to group the resistance, such as humidity, type of terrain, location, or terrain resistivity. The later one was shown in the previous studies to define different groups of resistances with reduced variance [19,20]. This will be also shown to be true latter in this work (see Fig. 1).

The remaining question is whether terrain resistivity can be guessed using available data, without resorting to *in situ* measures. Various machine learning techniques will be applied, including linear models, k-nearest neighbors, decision trees, and random forests [21]. If the resistivity groups were clearly defined, classificators could be trained using a cost function to model the impact of misclassification. However, in this first study, a direct regression will be used to estimate the resistivity of the terrain instead of only its group, in order to gain more insight into the prediction capabilities. The predictor could easily be specialized in a future implementation if the groups were fixed.

3 Results

A sample of 1085 measures was used in this first study to validate the proposed methodology. This data includes the measure of the resistance of the TTs, as well as information of the terrain resistivity, humidity, type of terrain, location, and model of TT, as well as other less relevant attributes.

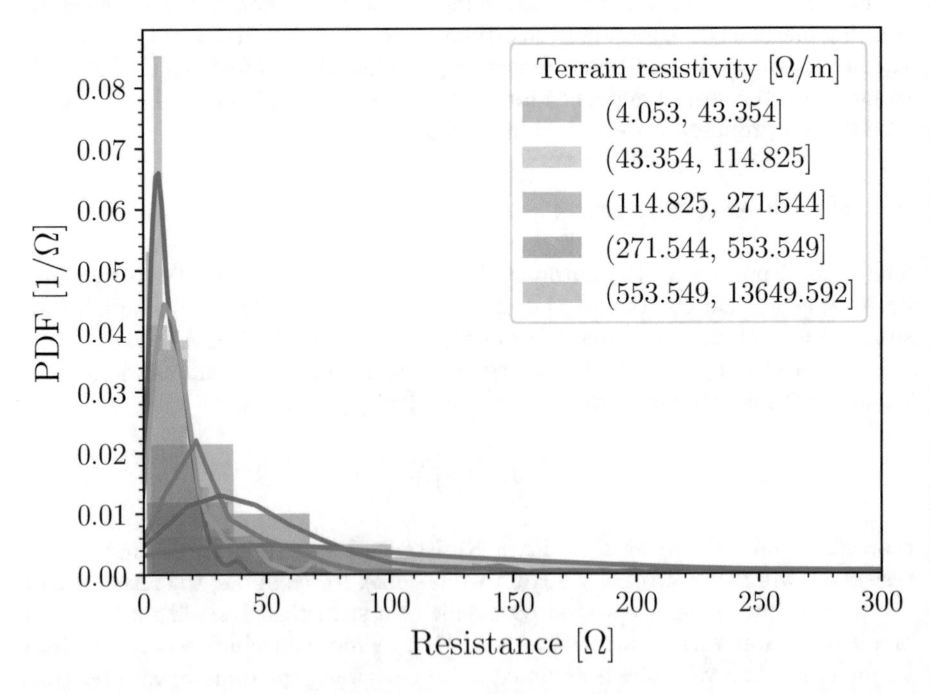

Fig. 1. Probability density function for each of the groups of terrain resistance (legend). Histograms are shown as the shaded area, while a kernel density estimation is added as the continuous line to aid visual inspection.

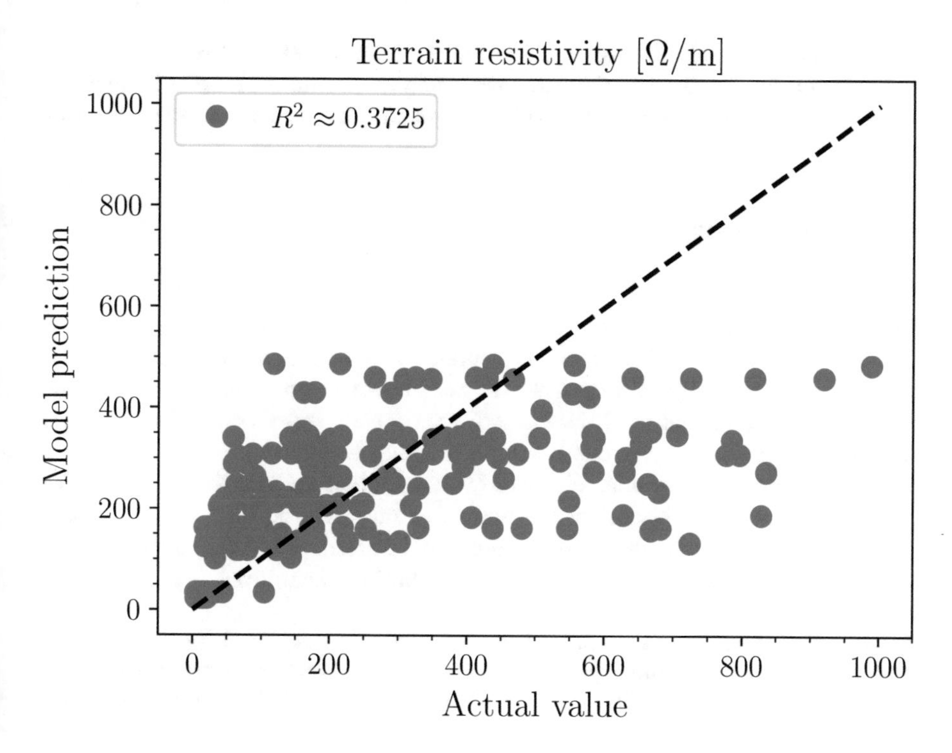

Fig. 2. Prediction of a random forest (y-axis) and actual value (x-axis) of terrain resistivity for the data in the test set. The predictor was built using a small set of attributes, as a proof of concept.

Figure 1 shows an estimation of the probability density function (PDF) of the resistance in the groups made by quintiles of resistivity terrain. The difference among the series shows a clear dependence on the terrain resistivity, which means that statistical conclusions will be different for each of the groups: TTs in terrains with less resistivity will require less sampling to provide a conclusion.

The number of quantiles shown here was chosen as 5 to allow extraction of statistically acceptable bounds for the first three groups in the studied sample, but this number could be changed in the future, as described before.

Prediction of terrain resistivity was addressed by the machine learning methods specified in Sect. 2. Best results are provided by the random forests, which achieve a mean coefficient of determination of $R^2 \approx 0.3075$ in a 3-fold cross-validation process. This value means that, although the method reduces the variance of the constant predictor and is, thus, a better criterion for sampling that a random procedure, there is still a fraction of the original variance to explain, probably by means of attributes not available in the dataset. A predicted vs. observed diagram is shown in Fig. 2.

Despite the small number of parameters used, the techniques allow for automated pattern search in the data, providing a method to guide the choice of TTs to find those probably fitting in the group we desire to study. Extended geological information to aid the determination of the terrain resistivity is currently being retrieved, which could considerably improve the prediction capabilities of the system.

4 Conclusions

A system integrating machine learning and statistical techniques for automatic optimization of the workflow of expert workers for the measure of the resistance of transmission towers has been implemented. The proposed system is based on a stratified sampling process, which was shown in previous works to reduce the error in the statistical inference. Estimation of the terrain resistivity was addressed by means of various machine learning methods, with random forest being the most accurate one. While only very limited information of the terrain was available, the methodology was shown to improve the simple random sampling. Future work will increase the number of attributes used to describe the terrain, allowing for improvement of the prediction capabilities.

Acknowledgments. This research has been partially supported by the European Regional Development Fund (FEDER) within the framework of the Interreg program V-A Spain-Portugal 2014–2020 (PocTep) under the IOTEC project grant 0123 IOTEC 3 E.

References

1. Singh, J., Gandhi, K., Kapoor, M., Dwivedi, A.: New approaches for live wire maintenance of transmission lines. MIT Int. J. Electr. Instrum. Eng. **3**(2), 67–71 (2013)
2. Gonçalves, R.S., Carvalho, J.C.M.: Review and latest trends in mobile robots used on power transmission lines. Int. J. Adv. Rob. Syst. **10**(12), 408 (2013)
3. Eltawil, M.A., Zhao, Z.: Grid-connected photovoltaic power systems: technical and potential problems—a review. Renew. Sustain. Energy Rev. **14**(1), 112–129 (2010)
4. Swanson, L.: Linking maintenance strategies to performance. Int. J. Prod. Econ. **70**(3), 237–244 (2001)
5. Ghazvini, M.A.F., Morais, H., Vale, Z.: Coordination between mid-term maintenance outage decisions and short-term security-constrained scheduling in smart distribution systems. Appl. Energy **96**, 281–291 (2012)
6. Smith, C.A., Corripio, A.B., Basurto, S.D.M.: Control automático de procesos: teoría y práctica. Number 968-18-3791-6. 01-A3 LU. AL-PCS. 1. Limusa (1991)
7. Na, M.G.: Auto-tuned PID controller using a model predictive control method for the steam generator water level. IEEE Trans. Nucl. Sci. **48**(5), 1664–1671 (2001)
8. Krishnanand, K.R., Dash, P.K., Naeem, M.H.: Detection, classification, and location of faults in power transmission lines. Int. J. Electr. Power Energy Syst. **67**, 76–86 (2015)
9. Taher, S.A., Sadeghkhani, I.: Estimation of magnitude and time duration of temporary overvoltages using ANN in transmission lines during power system restoration. Simul. Model. Pract. Theory **18**(6), 787–805 (2010)

10. Higgins, L.R., Mobley, R.K., Smith, R., et al.: Maintenance Engineering Handbook. McGraw-Hill, New York (2002)
11. Do, P., Voisin, A., Levrat, E., Iung, B.: A proactive condition-based maintenance strategy with both perfect and imperfect maintenance actions. Reliab. Eng. Syst. Saf. **133**, 22–32 (2015)
12. Zarnani, A., Musilek, P., Shi, X., Ke, X., He, H., Greiner, R.: Learning to predict ice accretion on electric power lines. Eng. Appl. Artif. Intell. **25**(3), 609–617 (2012)
13. Zhou, D., Zhang, H., Weng, S.: A novel prognostic model of performance degradation trend for power machinery maintenance. Energy **78**, 740–746 (2014)
14. De Faria, H., Costa, J.G.S., Olivas, J.L.M.: A review of monitoring methods for predictive maintenance of electric power transformers based on dissolved gas analysis. Renew. Sustain. Energy Rev. **46**, 201–209 (2015)
15. Trappey, A.J.C., Trappey, C.V., Ma, L., Chang, J.C.M.: Intelligent engineering asset management system for power transformer maintenance decision supports under various operating conditions. Comput. Ind. Eng. **84**, 3–11 (2015)
16. Weibull, W.: Wide applicability. Int. J. Appl. Mech. **103**(730), 293–297 (1951)
17. Hollander, M., Wolfe, D.A., Chicken, E.: Nonparametric Statistical Methods, vol. 751. Wiley, Hoboken (2013)
18. Chakraborti, S., Li, J.: Confidence interval estimation of a normal percentile. Am. Stat. **61**(4), 331–336 (2007)
19. Chamoso, P., De La Prieta, F., Villarrubia, G.: Intelligent system to control electric power distribution networks. DCAIJ Adv. Distrib. Comput. Artif. Intell. J. **4**(4), 1–8 (2015)
20. Chamoso, P., De Paz, J.F., Bajo, J., Villarrubia, G.: Intelligent control of energy distribution networks. In: International Conference on Practical Applications of Agents and Multi-Agent Systems, pp. 99–107. Springer (2016)
21. Pedregosa, F., Varoquaux, G., Gramfort, A., Michel, V., Thirion, B., Grisel, O., Blondel, M., Prettenhofer, P., Weiss, R., Dubourg, V., et al.: Scikit-learn: machine learning in Python. J. Mach. Learn. Res. **12**, 2825–2830 (2011)

Internet Data Extraction and Analysis for Profile Generation

Álvaro Bartolomé(✉), David García-Retuerta, Francisco Pinto-Santos, and Pablo Chamoso

BISITE Research Group, University of Salamanca, Calle Espejo s/n, 37007 Salamanca, Spain
{alvarob96,dvid,franpintosantos,chamoso}@usal.es

Abstract. Almost everything is stored on the Internet nowadays, and relying data on the Internet has become usual over the last years, directly increasing the value of data retrieval. Via Internet, data scientist can now find a way to access all the available data that is stored on the Internet, so they can turn that data into useful information. As people rely a lot of data on the Internet, they sometimes ignore the fact that all that data can be easily extracted, even when people think their information is safe or unavailable. In this article, we propose a system in where some data extraction techniques are going to be analysed in order to have an overview of the amount of data of a person that can be extracted from the Internet, and how that data is turned into information with an additional value in order to make data useful. The proposed system is going to be capable of retrieving huge loads of data from a person and process it using Artificial Intelligence, in order to classify its content to generate a personal profile containing all the information once its analysed. This research is based on personal profile generation of people from Spain, but it could be implemented for any other country. The proposed system has been implemented and tested on different people, and the results were quite satisfactory.

Keywords: Information recovery · Information fusion · Big Data · Profile generation

1 Introduction

Nowadays the loads of data indexed and stored on the Internet are huge, so it is, that Big Data techniques are getting more useful as the data to retrieve has a lot of relevance. The data from the Internet is useless until someone decides to use it and turn it into information, with a real value and meaning.

All the data stored on the Internet is sometimes public, which causes people to start distrusting relying their personal information on the Internet. Some search engines as Google were forced to implement mechanisms in multiple countries in order to offer people the possibility of exercising their "right to forget", whenever their information was exposed.

P. Novais et al. (Eds.): ISAmI 2019, AISC 1006, pp. 112–119, 2020.
https://doi.org/10.1007/978-3-030-24097-4_14

This article contains a research work that aims to retrieve all available information on the Internet in order to generate a personal profile of the person searched on the system. The information retrieval is based on multiple websites crawling and scraping that take as input parameters the full name and some keywords (relevant or related words with the person to generate a personal profile for), retrieving all the public and accessible data indexed on the Internet through search engines such as Google or Bing.

The personal profile is going to be based on information retrieval using Big Data ETL (Extract, Transform and Load) methodologies mainly via Web Crawlers and Web Scrapers, as Olston and Najork [1] explained on their survey. In order to classify the retrieved content we will be using AI techniques such as clustering [2] or text analytics [3] -in Spanish in this work, but extendable to any other language-, with the purpose of generating a valid and reliable personal profile.

The rest of the article is structured in such a way that Sect. 2 presents the background related to the present work. Section 3 presents an overview of the methodology used and the proposed system. Section 4 shows the results of the developed system. Finally, Sect. 5 shows the conclusions obtained with the system and the future lines of work.

2 Background

Nowadays, information retrieval and analysis from data stored on the Internet is more and more valuable due to its impact and relevancy, as people entrust a lot of information on the Internet.

Works like the one presented in [4] make it possible to speed up the search and retrieval of information from the Internet thanks to distributed architectures. In the same way, the emergence of technologies specifically aimed at both structured and unstructured distributed storage, such as NoSQL [5] or even HDFS (Hadoop Distributed File System), which have been successfully used to process large volumes of textual information [6], has also favoured this line of research.

The combination of these techniques together with classic AI methodologies, such as K-Means, Term Frequency - Inverse Document Frequency (TF-IDF), Support Vector Machine (SVM), Artificial Neural Networks (ANNs), Histogram of Oriented Gradients (HOG) have made it possible to develop the work proposed in this article.

These techniques are already being used in similar systems, related to the retrieval of information from the Internet, applied to different aspects. For example, K-Means is applied in [7] to create clusters to represent textual retrieved information in word embedded vectors or in works oriented to image retrieval, such as the one presented in [8].

TF-IDF is a very common technique applied to analyse the frequency of the terms in different texts (such as documents or webpages), which has been used in a lot of works, such as [9] or [10]. More specifically, it is a numerical statistic that is intended to reflect how important a term is to a document. In the work presented in this article, the use of this methodology helps to determine whether the keywords associated with a profile are sufficiently present in a document or web page to ensure that the text is related to the profile searched for.

SVM is a supervised learning model with associated learning algorithms that analyze data used for classification and regression analysis. It is very useful to search for facial patterns that can be matched, so that you can find more information about the profile you are looking for by identifying those patterns. This methodology has been used in works such as [11]. Similarly, HOG algorithms [12] or ANNs and their variations, such as Convolutional Neural Networks (CNNs) are also widely used today for facial recognition [13].

3 Proposed System

In this section the platform basis is explained and the main engine for the data extraction system, where all the data is collected from the Internet for its further analysis, as shown in Fig. 1.

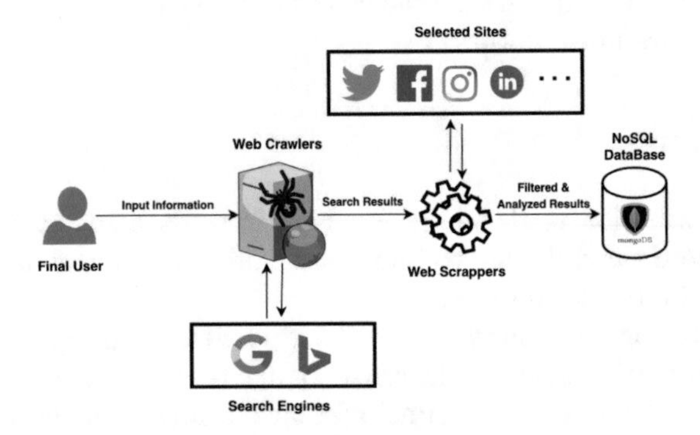

Fig. 1. Information retrieval methodology.

The platform is mainly based on Web Crawling and Web Scraping with the objective of retrieving personal information indexed and stored on the Internet directly or indirectly related to the person the user is looking for, for its further analysis in order to generate personal profiles. Therefore, the input parameters to the platform should be the know information of the person the user is looking for, such as the name, the surname or surnames and any related keyword that allows the system to single that person out, as it is like a distinctive factor in case we face the problem of two or more similar profiles. Thus, the output of the crawlers and the scrapers would be all the information that the system was able to retrieve and filter from the person-to-search, which is going to be inserted into a NoSQL database. Then the output of the data extraction module will be retrieved by the AI modules in order to analyse and clear all the information. Afterwards, once this process is completed, the analysed and filtered information will be shown in the dashboard, as a reply to the user's request.

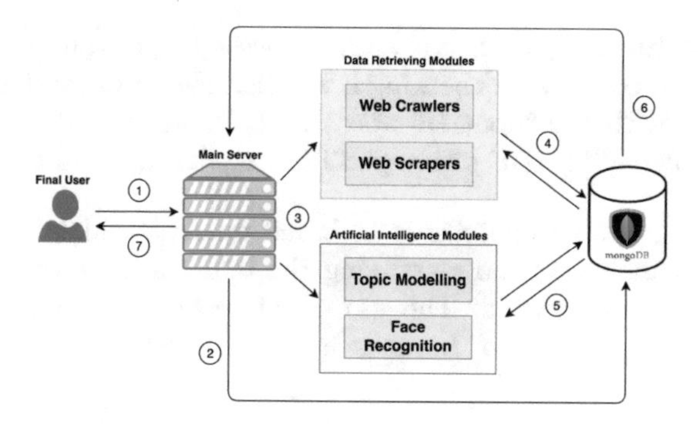

Fig. 2. Architecture of the proposed system.

As shown in Fig. 2, the whole platform is based on two main modules: the first one is the data retrieving module, where all the data stored on the Internet is collected, filtered, classified and inserted into a NoSQL database; and the second one is the AI module, where the retrieved data is analysed and classified.

Hence, the first group of modules is the one in charge of the ETL (Extract, Transform, and Load) process which implies retrieving the information that is extracted from the Internet via web crawlers and web scrapers, its later analysis and transformation so that it suits our proposed data model and the final load of dumped data on the database of our choice.

Web Crawler. A Web Crawler (often shortened to crawler) is an Internet Bot for bulk downloading web pages typically with the purpose of web indexing, but beyond that, there are a lot of different purposes for which a crawler can be used, like in this use case for retrieving information from the Internet to generate personal profiles. In the proposed system the main use of web crawlers is to automate searches and retrieve the HTML code from a website, as described in Fig. 1.

Web Scraping. Web Scraping consists on retrieving data from selected websites, so it involves fetching a web page and extracting data from its HTML code. It apparently is a form of copying, in which specific information is gathered and stored in a database, in this use case we decided to use a NoSQL database, MongoDB as explained before. All the retrieved webs via crawling are scraped, but this can also be useful when we have reliable sources from where we can get information that we know it is profitable by the time we generate the personal profile of the person-to-search.

Once we retrieve all the information possible that fits our data model, structured as a personal profile with fields such as name, surname/s, address, phone number/s, email, etc., we need to filter it. We use a filtering system when it comes to data classification based on:

1. **Regular Expression's** use is intended for retrieving just the information that matches, so e.g. we can check if a website contains an email in its HTML via regular expression, and the string that matches is classified as an email.

2. **Keyword Checking** is the very last process before dumping the data on MongoDB and consists on checking if the classified information has any kind of similarity with the input keywords and the generated keywords, as the result of applying TF-IDF to every HTML retrieved, from that profile.

The second group of modules is in based on Artificial Intelligence and so on in charge of analyzing and classifying the information either it is for topic modelling or face recognition. The use of AI techniques supposes a big step forward on profile generation, because it provides an additional value to the retrieved data.

Topic Modelling. Topic modelling consists on generating clusters with a related topic formed by some keywords. This AI process is done in order to classify information in clusters which include the most alike texts. In this use case, we will classify the related links retrieved from a profile search into categories, so we can sort out which collection of webs corresponds to each category. TF-IDF is used to vectorize the corpus, and k-means to define the clusters.

Face Recognition. Face Recognition consists on training an AI module with face landmarks of the person to classify images from and select just the ones that match the input image, the real and verified image of the person-to-search. As we retrieve a lot of image links related to a person, they need to be classified to keep just the image links where the person-to-search appears, so we pass the first image retrieved by Google image search, of the person-to-search, as an input to the face recognition module. In depth at a lower level we use a CNN to extract 128 features of each faces, which are later on compared to other faces features [14]. Preprocessing is a compulsory step in order to obtain a common framework for all images, it applies gradient-based algorithms and affine transformations. Faces comparison is made using SVM.

Another important issue when developing the proposed system was to use a fast access database in order to speed up read/write operations, due to we are facing huge loads of data. So on, we decided to use a NoSQL database, MongoDB, because it has a scheme-less structure.

4 Overview of the Results

This section presents the results obtained with the implementation of the proposed system. The way we represent the data of the generated profile is really important to show the user all the information already retrieved and analysed, so the user can see that information as a portfolio of the generated profile.

Hence, data visualization is made in a dashboard like the one presented in Fig. 3, where the user sends an input containing the known information of the person-to-search, being at least the full name and one keyword mandatory, which is later processed by the proposed system. When the proposed system ends, a MongoDB document containing all the information from the person-to-search is retrieved and shown in the dashboard.

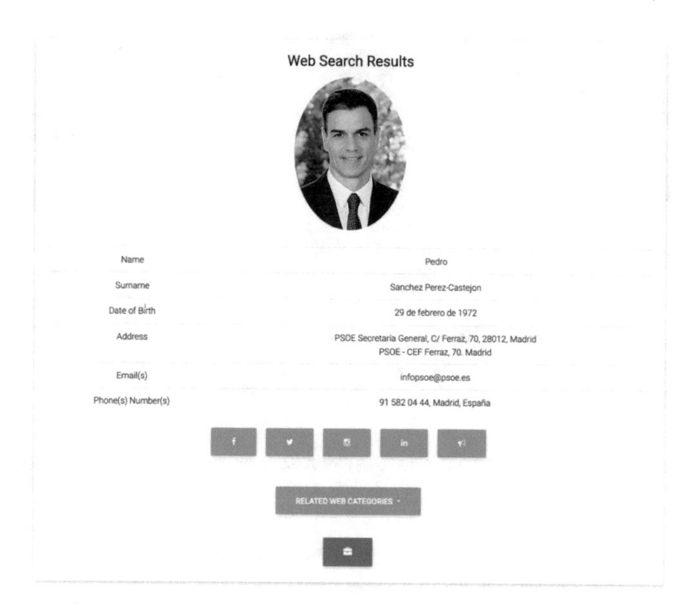

Fig. 3. Dashboard panel for data visualization.

Being able to generate a personal profile based on the indexed data on the Internet, can be a quite hard task to carry out. We have faced some problems when it comes to filtering information, because all the retrieved content does not match the personal profile of the person-to-search. With the purpose of keeping just useful data we developed a filtering system based on keywords, where the input keywords and the generated ones from TF-IDF analysis are combined so we can check coincidences between the name of the person-to-search and the keywords, and if it does not match we discard it, otherwise we keep it.

In order to avoid automated search limits on Google or Bing, we developed a rotating system based on modifying the headers when sending requests to both search engines changing the User-Agent used to process the request and a proxy system to avoid IP limitations. Content indexed on Bing is more scarce than Google's, so when what we seek it to generate a personal profile, what we need is a search engine from where we can get as much data as we can in the shortest time possible, so this is the main reason why we chose Google over Bing as plotted in Fig. 4.

The accuracy of the proposed system was pretty impressive overall, as seen in Fig. 3 all the fields are correctly fulfilled meaning that filtering techniques cleaned the original information and kept the useful one, with a error of over-information sometimes. The conclusion as to the accuracy of the system was that the relevance of input keywords had much importance on the results, because keywords are the key when it comes to filtering the data once it is retrieved, before dumping it on the database. As this was said, the more accurate keywords the better results; but keyword relevance is determined by the quality of the

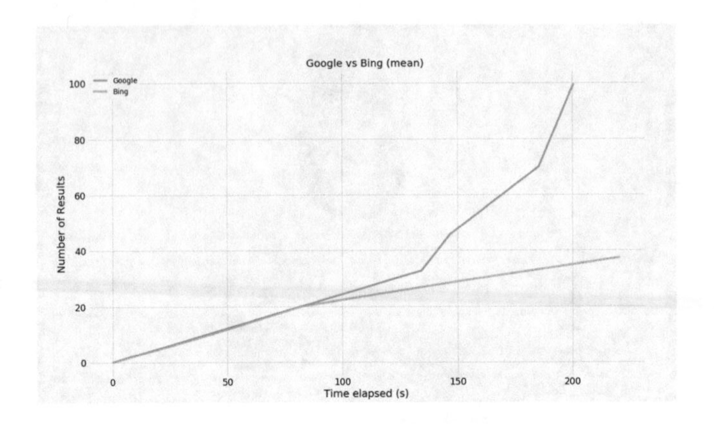

Fig. 4. Statistic comparison between search engines performance.

keywords (their real relevance on the person-to-search) not by the amount of them, so the more keywords the less accurate the system is.

5 Conclusion and Future Work

To conclude, our research let us know which are the most useful and reliable tools when crawling and scraping websites in order to retrieve data from them. So on we developed a system with the objective of retrieving all the information indexed on the Internet of a person, termed as person-to-search, and after analysing and filtering all the information, showing it in a dashboard. As already explained, combining Data Extraction techniques, like Web Crawling and Web Scraping, with AI algorithms is a breakthrough on profile generation based on data extracted from the Internet.

As a future work, our aim is to migrate the developed system to a commercial cloud environment with no (practical) limit of computing capacity. Our goal is to release the tool for public use, although always in a limited way, i.e. only a request per IP address and hour, in order to be able to control the use of the system. If, once launched, someone wants to make greater use of the system for research purposes, they will have to contact the authors. In the same way, once the tool is launched, it will continue to incorporate functionality to keep it updated.

Acknowledgements. This research has been partially supported by the European Regional Development Fund (FEDER) within the framework of the Interreg program V-A Spain-Portugal 2014–2020 (PocTep) under the IOTEC project grant 0123 IOTEC 3 E.

References

1. Olston, C., Najork, M., et al.: Web crawling. Found. Trends® Inf. Retrieval **4**(3), 175–246 (2010)
2. Allahyari, M., Pouriyeh, S., Assefi, M., Safaei, S., Trippe, E.D., Gutierrez, J.B., Kochut, K.: A brief survey of text mining: classification, clustering and extraction techniques. arXiv preprint arXiv:1707.02919 (2017)
3. Moreno, A., Redondo, T.: Text analytics: the convergence of big data and artificial intelligence. IJIMAI **3**(6), 57–64 (2016)
4. Bahrami, M., Singhal, M., Zhuang, Z.: A cloud-based web crawler architecture. In: 2015 18th International Conference on Intelligence in Next Generation Networks, pp. 216–223. IEEE (2015)
5. Jose, B., Abraham, S.: Exploring the merits of NoSQL: a study based on MongoDB. In: 2017 International Conference on Networks & Advances in Computational Technologies (NetACT), pp. 266–271. IEEE (2017)
6. Sun, S., Gong, J., Zomaya, A.Y., Wu, A.: A distributed incremental information acquisition model for large-scale text data. Cluster Comput. **20**, 1–12 (2017)
7. Roy, D., Ganguly, D., Mitra, M., Jones, G.J.F.: Representing documents and queries as sets of word embedded vectors for information retrieval. arXiv preprint arXiv:1606.07869 (2016)
8. Ali, N., Bajwa, K.B., Sablatnig, R., Mehmood, Z.: Image retrieval by addition of spatial information based on histograms of triangular regions. Comput. Electr. Eng. **54**, 539–550 (2016)
9. Rivas, A., Martín, L., Sittón, I., Chamoso, P., Martín-Limorti, J.J., Prieto, J., González-Briones, A.: Semantic analysis system for industry 4.0. In: International Conference on Knowledge Management in Organizations, pp. 537–548. Springer (2018)
10. Binkheder, S., Wu, H.-Y., Quinney, S., Li, L.: Analyzing patterns of literature-based phenotyping definitions for text mining applications. In: 2018 IEEE International Conference on Healthcare Informatics (ICHI), pp. 374–376. IEEE (2018)
11. Shah, J.H., Sharif, M., Yasmin, M., Fernandes, S.L.: Facial expressions classification and false label reduction using LDA and threefold SVM. Pattern Recogn. Lett. (2017)
12. Dalal, N., Triggs, B.: Histograms of oriented gradients for human detection. In: IEEE Computer Society Conference on Computer Vision and Pattern Recognition, CVPR 2005, vol. 1, pp. 886–893. IEEE (2005)
13. Kasar, M.M., Bhattacharyya, D., Kim, T.H.: Face recognition using neural network: a review. Int. J. Secur. Appl. **10**(3), 81–100 (2016)
14. Amos, B., Ludwiczuk, B., Satyanarayanan, M., et al.: OpenFace: a general-purpose face recognition library with mobile applications. CMU School of Computer Science (2016)

Original Content Verification Using Hash-Based Video Analysis

David García-Retuerta[1](✉), Álvaro Bartolomé[1], Pablo Chamoso[1], Juan M. Corchado[1,2,3], and Alfonso González-Briones[1]

[1] BISITE Digital Innovation Hub, University of Salamanca, Edificio Multiusos I+D+i, Calle Espejo 2, 37007 Salamanca, Spain
{dvid,alvarob96,chamoso,corchado,alfonsogb}@usal.es

[2] Department of Electronics, Information and Communication, Faculty of Engineering, Osaka Institute of Technology, Osaka 535-8585, Japan

[3] Pusat Komputeran dan Informatik, Universiti Malaysia Kelantan, Karung Berkunci 36, Pengkaan Chepa, 16100 Kota Bharu, Kelantan, Malaysia
https://bisite.usal.es/en

Abstract. Nowadays, the internet makes it possible for us to upload and share content online. However, the problem is that copying online content has become very easy and has put copyright content at risk. State-of-the-art tools have been designed to detect plagiarised images or texts through the detection of similarities in them. However, there has not yet been a tool for the identification of plagiarised videos which are made up of the fragments of other original videos, that may be legally protected by their authors. This paper presents a tool that has been developed to identify videos created from the fragments of other existing content. The system has been evaluated using videos from the World Cup held in Russia in 2018, some had original content while others were made up of copied fragments. In this way we have been able to verify the feasibility of the system in correctly matching original videos with the plagiarised ones. The results have been satisfactory.

Keywords: Video processing · Image matching · Big data · Software tool

1 Introduction

The detection of original sources of a video is an important step in many technological applications. Watermarks have been used for detecting the authorship of videos in the past [1], however this affects the quality of the content negatively. Modern copyright protection methods do not alter the frames, making it easier to detect content with copyrights.

In addition, big technological companies like Facebook and YouTube are believed to have internal hash-based systems for the detection and automatic deletion of terrorist propaganda. Counter-terrorist agencies should also take

P. Novais et al. (Eds.): ISAmI 2019, AISC 1006, pp. 120–127, 2020.
https://doi.org/10.1007/978-3-030-24097-4_15

advantage of such systems, as the extremist projects detected by researchers had done between 70% and 80% of their networking online [2].

This paper combines web scrapping and video comparison for the identification of the original sources of a video automatically. Having introduced the topic of this paper briefly, the next seven sections further elaborate on it.

2 State of the Art

Several methods for detecting the original sources of video fragments have been proposed previously. Most of them belong to one of the following two categories, or a mixture of both: traditional approaches and Deep Learning methods.

Traditional Approaches. Fingerprint extraction is a common technique for the detection of copied videos [1]. Cryptographic algorithms such as SHA and MD5 [3] perform very well in detecting copied videos but they are not suitable for the detection of old plagiarised videos, as they are incapable of detecting common video editing. Content based hashing is another approach which extracts faces and other landmarks or features for future comparison [1,4,5]. Comparative studies of local feature descriptors have shown LBP histograms to perform well in reduced datasets [6–8]. Recent studies have evaluated local SIFT descriptors against more recent methods which achieved a great improvement, making SIFT descriptors obsolete [9].

Deep Learning Methods. Jiang *et al.* [9] have evaluated the performance of CNN and their own developed SCNN (Siamese Convolutional Neural Network). A standard Convolutional Neural Network (AlexNet) provided very accurate results, although the processing time is significantly higher than with the traditional methods. Although the processing time of SCNN is greatly reduced its results are less accurate. Thus, the SCNN has a lot of room for improvement.

Mixed Methods. Researches have suggested the use of Deep Learning for the generation of hash-like representations of frames [10], resulting in high accuracy. However, processing time is increased due to the mixture of very different techniques. A representation of such a architecture can be found below (Fig.1).

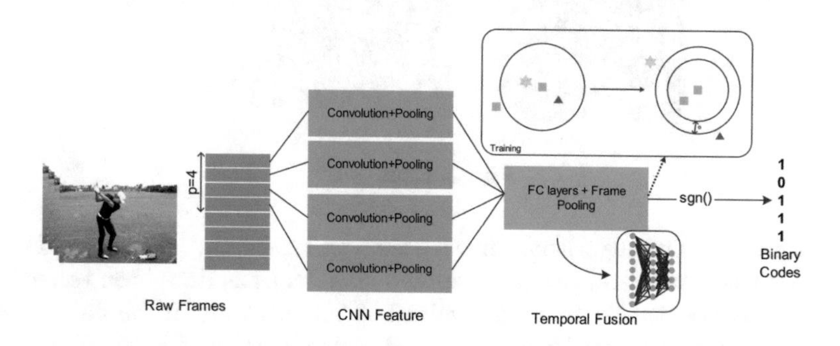

Fig. 1. Mixed method for video search [10].

3 Overview of the Method

This section describes the frame comparison process, which is summarised in Fig. 2. Information about algorithm implementation can be found in Sect. 6. The original content of each frame has been stored excluding irrelevant details. New frames are compared with the basic information stored in the database. If the differences between them are below an established threshold, the result is a match.

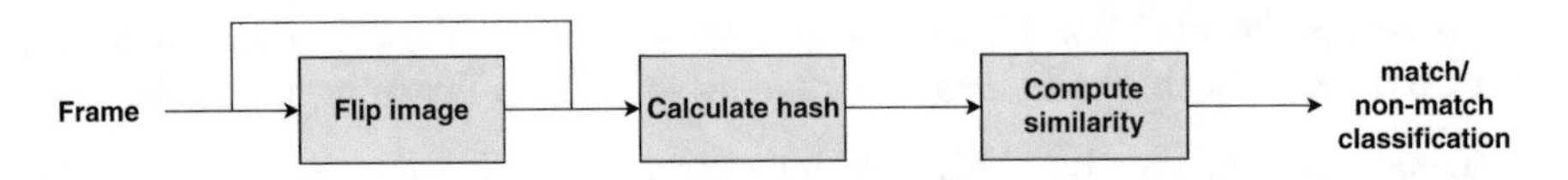

Fig. 2. Method preview.

4 Data Sets and Methodology

Datasets. A set of 29 videos have been used with original content from the 2018 FIFA World Cup in Russia. The total size of the videos is 104 GB and they contain 4,958,718 frames. Each video lasts either 1 h and 50 min or 2 h and 45 min. The new video is a summary of the best goals in the World Cup whose length is 5 min and contains 7,631 frames. The new frames have been altered in colour, brightness, watermarks, scoreboards, logos, resolution and aspect ratio. Some examples can be found in Fig. 3.

Fig. 3. Example.

Methodology. The algorithms have been applied to the videos varying the frame extraction rate and several threshold combinations have been tested. The original content database is fulfilled with 980,000 elements in the final version. Python has been used for implementing the algorithms and MongoDB for the databases. All the tests have been carried out with an i7-7700K processor.

5 Architectural Framework

In this section we are going to break down our architectural framework as shown in Fig. 4, explaining each of the ETL (Extract, Transform and Load) processes in video retrieval and storage. The use of ETL techniques based on Big Data is an important element due to the high load of information when retrieving and storing videos indexed on the Internet from the 2018 FIFA World Cup held in Russia.

The main focus of the case study was to compare the feasibility of different search engines, like Google, Bing or DuckDuckGo, in the retrieval of indexed videos. After some research we noticed that there were a lot of similar videos that had been recorded from different perspectives or has been produced by different TV channels, most videos focused on goals or important parts of the matches. Once we retrieved up to 1,000 videos, we gained a better understanding of the quality of each of the analysed search engines, arriving at the conclusion that the Google search engine performs better than Bing or DuckDuckGo.

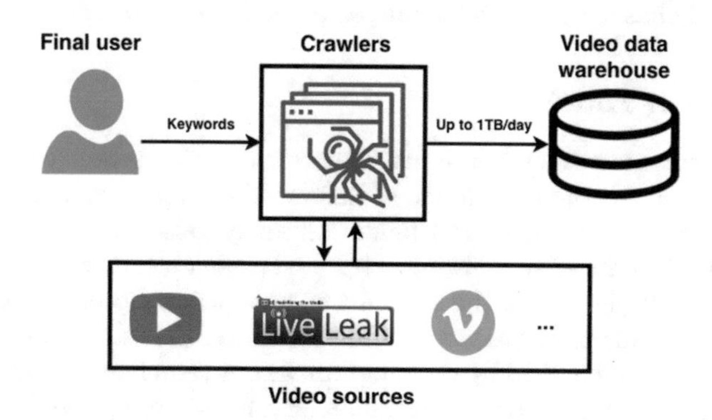

Fig. 4. Proposed system for storage of retrieved videos.

The techniques leveraged in the ETL process include:

- **Extract:** the entry point of this process are the keywords thanks to which instances of the video retrieving web crawlers are launched to retrieve the raw videos associated with a given set of keywords. In this paper, the proposed system retrieved videos from the 2018 FIFA World Cup held in Russia. Once we recover all the videos from the data warehouse we need to transform and store them.
- **Transform:** this process entails transforming the raw data of the retrieved videos into the desired output. Hence, we must parse and decompose each video into frames, which can then be converted into a hash.
- **Load:** thanks to the schemaless structure and fast access to read/write operations of the MongoDB the frame of every video is inserted into it as a hash. Once we have every video transformed and inserted into our MongoDB, we can proceed to the next step which is the analysis of each frame.

6 Implementation and Case Study

The implementation of the proposed system is described in this section and a case study is performed in which we evaluate different options and their effect on the system. Throughout this section, the algorithm configuration is: frame extraction rates of 5Fr/s and 3Fr/s respectively; threshold of 4 units.

6.1 Choice of Optimal Frame Extraction Rate

The aim of our experiments was to determine the amount of frames that should be extracted per second in order to achieve best performance. A small amount of frames will make the system ineffective in matching original videos with the copied ones, while too many frames may cause false positives. Thus, it was essential to identify the optimal number of frames. Our tests showed that the best performance is achieved when extracting 5 frames/sec - 3 frames/sec rate for the plagiarised and original videos, which resulted in 33040 and 916 frames, respectively. It has a 1.214% error rate at detecting copied videos.

6.2 Choice of Threshold

To ensure correct identification of matching videos it is necessary to established a suitable threshold; the percentage of difference between two matching frames must not surpass an established maximum. This may cause some false positives but its probability is reduced with frame extraction at an optimal rate. Furthermore, it is possible to implement in the system a complementary program which would thoroughly check all suspicious videos. Videos with less than three matching frames should by default be checked by the program, as our tests have shown that videos with false positives often have just one matching frame, exceptionally two.

We use a threshold of 4 units in our tests (frames with up to 6.25% hash difference are considered as matching) which, in the majority of cases, have correctly matched the copied videos with their original sources. Tests have also been carried out with a threshold of 3 units (4.69% hash difference) where all the original sources were identified and no false positives were obtained. However, it caused the processing time to triple and as a result has not been studied any further.

6.3 Hash Generation

A pHash-based algorithm is used. Our process for generating the hash associated to a frame is as follows: Firstly, the average pixel value must be calculated. It will be used to transform the image into a binary number. Pixels above the average are transformed into 1 s and pixels below the average into 0 s.

6.4 Pre-processing Algorithm

The algorithm is divided into two sub-algorithms. The input of the first sub-algorithm is a video and its output a sequence of hashes. It works as follows:

Firstly, it extracts frames from a given video. Each frame is flipped so that the darkest half is on the left, it is resized into 16×16 pixels and converted into black and white. Afterwards, the associated hash is generated as described in Sect. 6.3. If the resulting hash is a constant, it is dismissed.

Thus, the algorithm reduces the size of the frames by 99.93% and it performs very well even though the frames are flipped. In our tests, the proposed algorithm pre-processed the videos in 3 h 41 min and 52 s.

6.5 Matching Algorithm

The input of the second sub-algorithm are the hash database and the hashes of a new video, its output is a list of matching frames. This algorithm decides whether new frames are original or not. It works as follows:

Firstly, it calculates the Hamming distance of each frame of the new video to each frame of the database. If the distance is lower than the defined threshold, both hashes are considered a copy. Otherwise, they are consider original.

When tested with the previous database and a 75.5 MB video, consisting of 7631 frames, the video was analysed in 34 min and 57 s. All the copied fragments of the video were detected and their original sources identified.

The basic idea behind the algorithm is that most of the irrelevant information of each frame is dismissed, only the most basic features are compared with each other. The basic features are slightly altered using common video editing methods, however, such frames are normally below the defined threshold.

6.6 Structure

Our implementation consists of two programs and two databases as we believe it to be easily implemented in real-world situations.

Program 1 implements the pre-processing algorithm and writes the hashes in database 1.

Program 2 takes as input the hashes stored in database 1 and a new video whose hashes are extracted using the pre-processing algorithm. Afterwards, it applies the matching algorithm to the corresponding frames and writes the matches in database 2.

Therefore, *database 1* stores our hashes, their respective video title and frame position. *Database 2* stores hash matches. It contains the title of the (original) video which had a match, and the positions of the frames which matched.

Figure 5 represents program 1 and the initial part of program 2.

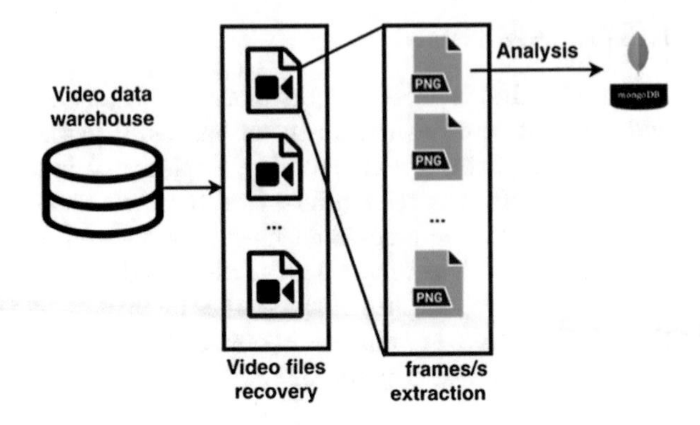

Fig. 5. Schema of the process used to convert videos into hashes.

7 Overview of the Results

The algorithm has demonstrated resistance to watermarks, scoreboards, colour filters and re-sized images. Therefore, the widely used methods of avoiding copyright detection cannot deceive this method, not even if several of them are applied at once.

The error rate of frame comparison is below $4 \cdot 10^{-8}\%$, which makes the algorithm suitable for video comparison.

Best performance was obtained with a threshold of 3 units and 5 frames per second - 12Fr/s extraction rate (in the original and new videos respectively). The best configuration for real-time video comparison is a threshold of 4 units, 3Fr/s – 5Fr/s extraction rate (Fig. 6).

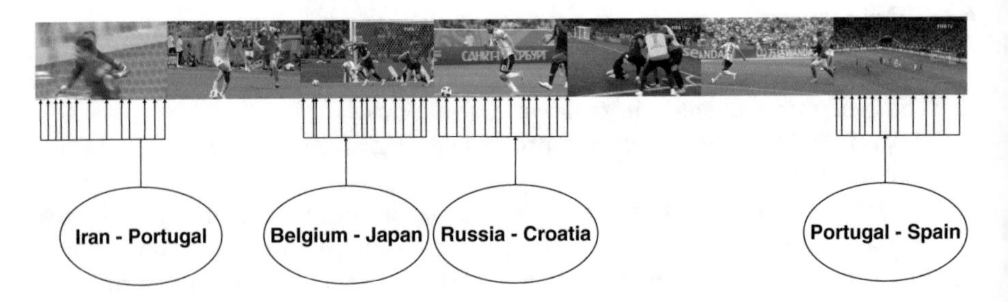

Fig. 6. Identified video sources.

8 Conclusions and Future Work

The original sources of a new video can be detected automatically and the robust performance of the proposed system make it suitable for a real-life analysis. Widespread video editing methods used to avoid copyright infringement are easily detected bu the proposed algorithm. This is a highly parallel algorithm as each individual frame comparison, processing and extraction can be performed at the same time in a multi-core processor. Therefore, it is highly scalable, making it suitable for big scale applications.

The future lines of research include the detection of cropped images and added borders. Another goal is to further improve the performance of the system, extending this method to greater scopes, and optimising it for GPU parallel computing.

Acknowledgements. This research has been partially supported by the European Regional Development Fund (FEDER) within the framework of the Interreg program V-A Spain-Portugal 2014-2020 (PocTep) under the IOTEC project grant 0123 IOTEC 3 E.

References

1. Lian, S., Nikolaidis, N., Sencar, H.T.: Content-based video copy detection–a survey. In: Intelligent Multimedia Analysis for Security Applications, pp. 253–273. Springer, Heidelberg (2010)
2. Koehler, D.: The radical online: individual radicalization processes and the role of the internet. J. Deradicalization **1**(1), 116–134 (2014)
3. Lian, S.: Multimedia Content Encryption: Techniques and Applications. Auerbach Publications, Boston (2008)
4. Gengembre, N., Berrani, S.-A.: The orange labs real time video copy detection system-TRECVID 2008 results. In: TRECVID (2008)
5. Gengembre, N., Berrani, S.-A.: A probabilistic framework for fusing frame-based searches within a video copy detection system. In: Proceedings of the 2008 International Conference on Content-based Image And Video Retrieval, pp. 211–220. ACM (2008)
6. Wu, X., Hauptmann, A.G., Ngo, C.-W.: Practical elimination of near-duplicates from web video search. In: Proceedings of the 15th ACM International Conference on Multimedia, pp. 218–227. ACM (2007)
7. Song, J., Yang, Y., Huang, Z., Shen, H.T., Hong, R.: Multiple feature hashing for real-time large scale near-duplicate video retrieval. In: Proceedings of the 19th ACM International Conference on Multimedia, pp. 423–432. ACM (2011)
8. Law-To, J., Chen, L., Joly, A., Laptev, I., Buisson, O., Gouet-Brunet, V., Boujemaa, N., Stentiford, F.: Video copy detection: a comparative study. In: Proceedings of the 6th ACM International Conference on Image and Video Retrieval, pp. 371–378. ACM (2007)
9. Jiang, Y.-G., Wang, J.: Partial copy detection in videos: a benchmark and an evaluation of popular methods. IEEE Trans. Big Data **2**(1), 32–42 (2016)
10. Liong, V.E., Lu, J., Tan, Y.-P., Zhou, J.: Deep video hashing. IEEE Trans. Multimed. **19**(6), 1209–1219 (2017)

ME^{3}CA - Monitoring Environment Exercise and Emotion by a Cognitive Assistant

J. A. Rincon[1], A. Costa[2], P. Novais[2], V. Julian[1(✉)], and C. Carrascosa[1]

[1] D. Sistemas Informáticos y Computación,
Universitat Politècnica de València, Valencia, Spain
{jrincon,vinglada,carrasco}@dsic.upv.es
[2] ALGORITMI Center, University of Minho, Braga, Portugal
{acosta,pjon}@di.uminho.pt

Abstract. The elderly population has increased dramatically in today's society. This fact implies the need to propose new policies of attention to this group but without increasing social spending. Currently, there is a need to promote the care of elderly people in their own homes, avoiding being transferred to saturated residences. Bearing this in mind, in recent years numerous approaches have tried to offer solutions in this sense using the continuous advances in new information and communication technologies. In this way, this article proposes the employment of a personal assistant to help the elderly in the development of their daily life activities. The proposed system, called ME^{3}CA, is a cognitive assistant that involves users in rehabilitating exercise, consisting of a sensorization platform and different integrated decision-making mechanisms. The system tries to plan and recommend activities to older people trying to improve their physical activity. In addition, in the decision making process the assistant takes into account the emotions of the user. In this way, the system is more personalized and emotionally intelligent.

1 Introduction

Worldwide population is ageing fast and the tendency is that elderly people (65+) surpass the number of births [1]. This means that in the near future society will have difficulties in provide social care to elderly people due to the lack of funds (due to a decay in the number of people that is available to contribute monetarily) and the lack of social/medical workers, as demonstrated in [2]. This means that families will have complications in care for the elders themselves or through care services [2].

Social services are currently unable to provide care for the all elderly people who need them and this issue will be even more critical in the future [3]. Adapting these services will have a high cost due to the demand of infrastructures and specialised personnel [4]. By 2060 44.4 Million people of over 65 years will need assistance on their activities of daily living (ADLs) [5].

P. Novais et al. (Eds.): ISAmI 2019, AISC 1006, pp. 128–135, 2020.
https://doi.org/10.1007/978-3-030-24097-4_16

Aggravating this social issue is the increasing number of elderly people with cognitive issues such as Alzheimer's, as stated in the study [6]. Caring for people with these cognitive issues require special training, which currently people that have this training is scarce [4]. Furthermore, the level of attention that these elders need (medication intake and medical control - absence of medication can rapidly worsen the health condition [7]) makes it impossible for caregivers to care for more than one person [5].

One solution may be adapting the elder's home with technological devices and services so they are able to stay longer at their homes comfortably and safely. Moreover, keeping the elderly at their homes has great benefits [8] as the environment is familiar, they have to be less moved (increased fall risk) and their relatives and neighbours are able to visit them often. But keeping them at home has also issues, like stagnation and doing repetitive and unchallenging tasks (which are often associated with cognitive problems like dementia [9]). To overcome this issues exercises and activities may be used to engage the elderly into exercise physically and cognitively.

The exercises should be meaningful (or useful) to the elderly, and should have a positive impact in their life [9]. Apart from this, the exercises have to be carefully monitored, as well as the elderly (through sensor systems) and the environment (for outside exercises/activities).

To this, some solutions have been designed [10], like the Buddy robot [11] that displays human-like emotions and is able to give information about specific tasks and maintain a conversation fluently. Or the InTouch Health [12] that is a robotic tele-doctor armed with sensors, thus overcoming the need of a doctor's physical presence. More oriented to exercises is the PHAROS project [13] that uses a Pepper Robot to teach and evaluate physical exercises providing this information to the caregivers. Lastly there is the PersonALL project [14] that monitors the elder's behaviour and displays health-related suggestions searching for any decay of motor abilities.

The issue with these projects is that they fail to address two important aspects: emotions and environment. Emotions influence the physical health (like pain), thus negative emotions have a great impact physically [15]. In terms of the environment, the previously mentioned projects do not consider (or suggest) outdoors exercises. Thus, if these types of exercises are suggested, the quality of the environment (e.g. torrid or cold weather, air pollution) must be measured and the features of the exercise adjusted.

In this paper we present **ME^{3}CA**, which goal is to use emotions and the quality of the environment to change the decision-making process. This means that the suggestions are personalised and an emotional bond is created due to the display of human-like emotions by the assistant.

This paper is structured in the following way: Sect. 2 presents the system description (detailing the hardware and software components); finally, Sect. 3 presents the conclusions and the future work proposals.

2 System Description

This section presents our system, which has been divided into hardware and software. The hardware is in charge of the acquisition of biological signals and environmental conditions and the localised software as an external service that allows to generate and adapt a sequence of personalised exercises for each individual.

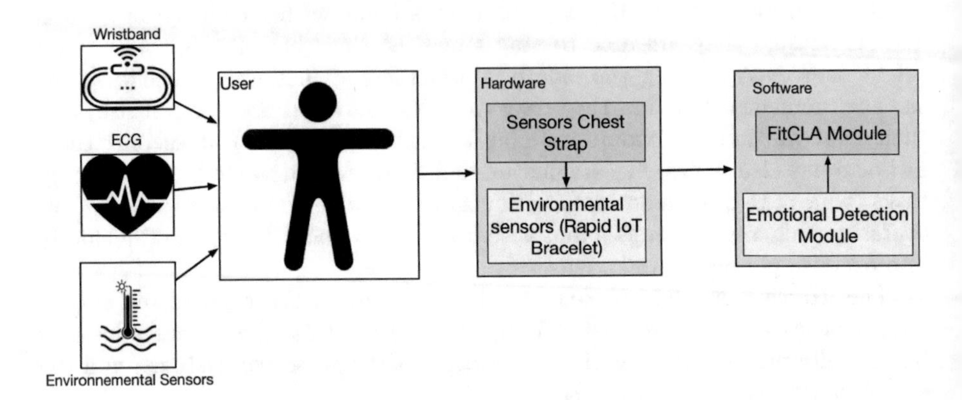

Fig. 1. System's components.

The hardware part has been divided into two systems. The first one is a set of biosensors that allows to capture a set of biosignals. These can be used to adapt the exercises dynamically, as well as to perceive the evolution of the emotion of the person. The second system is composed of a device that incorporates environmental sensors such as temperature, humidity, light intensity, air quality and CO2 levels. Using this information our system can adapt the sequence of exercises, so that the final experience obtained by each user of our system is the most appropriate sequence of exercises depending not only on the physical evolution of the person during the sequence of exercises, but also the evolution of their emotions.

Figure 1 shows the different components forming the ME^{3}CA system. This components can be classified in the following groups:

- User: this is the main part of the system as it is not only the source of the input sensor data, but also the goal of the actions carried out by the system.
- Hardware: This group is formed by all the different sensors that can be used to perceive the evolution of the person to the exercise sequence. In fact, this sensors are grouped in two different artefacts: the *Sensors Chest Strap*, formed by a set of sensors that will go in the user chest, and they even could be linked to a slim-fit exercise t-shirt; and the *Sensors Wristband* formed by a set of sensors collocated in a wristband weared by the user.

– Software: This group is formed by all the software modules in charge of calculating information and using this information to create and/or adapt the exercise sequence of a user according to his user profile and dynamic evolution. This modules are: *Empathy Module* according to the sensor information, calculates the current user's emotion; and the *FitCLA* which calculates the proper exercises sequence or adapts the current one according not only to the user profile, but also his current physical and emotional stress.

In the following subsections, the hardware and software parts of the ME^{3}CA system are detailed.

2.1 Hardware Description

This section presents the description of the hardware, which has been divided into two sub-systems. The first is composed of a chest belt, which incorporates a series of sensors. These allow us to acquire the signals of ECG (electrocardiography), EDA (electrical activity of the skin). The second is composed of a bracelet, which incorporates sensors of temperature, humidity, motion detector, fall detector, air quality and CO2 levels. The data acquired by the two systems are sent to a web service, to be analysed in depth. Using AI techniques such of Deep-Learning [16], or Neuronal Networks [17], it is possible to recommend the physical activities, as well as, the monitoring of the same one allowing the modification of the activity in real time.

To acquire these signals, the chest strap needs a communication interface between the skin and the capture device. This interface is achieved through electrodes, which are made of stainless steel.

The arrangement of the electrodes is equal to a triangle, this triangle is known as the *Einthoven* triangle (Fig. 2). It allows us to capture the standard bipolar leads, which are the classic electrocardiogram leads, recording the potential differences between the electrodes located at the different extremities.

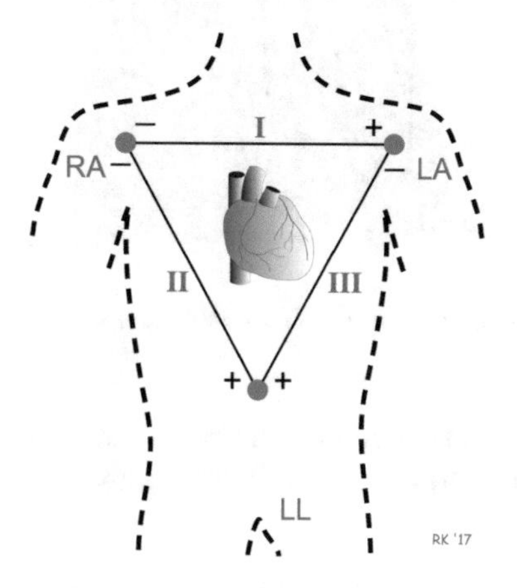

Fig. 2. *Einthoven* triangle.

D1 or I: Potential difference between right arm (RA) and left arm (RL). Its vector is in the 0° direction.

D2 or II: Potential difference between right arm (RA) and left leg (LL). Its vector is in the 60° direction.

D3 or III: Potential difference between left arm (RL) and left leg (LL). Its vector is in the 120° direction.

The data acquired by the chest belt is made using an Arduino 101. Using the analog-to-digital conversion pins, the system is able to convert the analog heart signal to a digital one. The ECG signal acquisition system has a sampling rate of 250 to 300 samples per second and uses a Butterworth [18] band-pass filter with 100 and 250 Hz cut-off frequencies. Just as a Butterworth filter rejects a 50 Hz band, this filter allows us to eliminate electrical noise.

This digital data is then used to calculate the beats per minute, as well as to make telemedicine applications. In turn, the chest belt has two communication systems. The first one is a Bluetooth communication system of low consumption or BLE (Bluetooth Low Energy) and a Near Field Communication (NFC) system. The Bluetooth communication system allows the chest belt to send the digitalized data to a smartphone in streaming, while the NFC system allows the transfer of such signals to other systems with NFC communication.

Fig. 3. Rapid IoT development Board.

The bracelet, on the other hand, has been developed using the Rapid IoT development system (Fig. 3) of the NXP[1] company.

[1] https://www.nxp.com/support/developer-resources/rapid-prototyping/nxp-rapid-iot-prototyping-kit:IOT-PROTOTYPING.

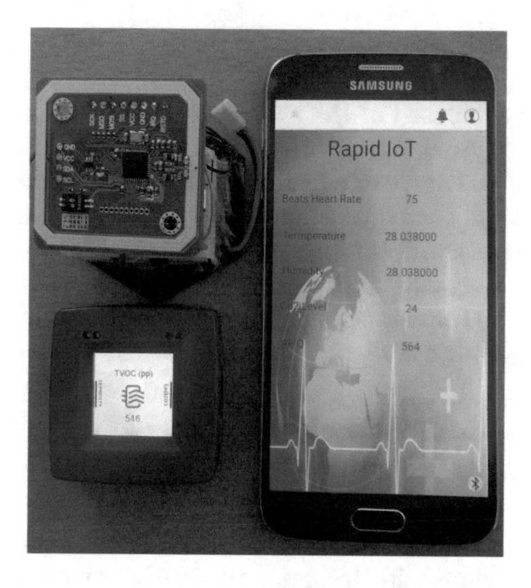

Fig. 4. A view of all the components: Chest Strap, bracelet and smartphone APP.

A new app allowing to connect the bracelet to an smartphone has been created using this development system (Fig. 4). This app allows to acquire and to observe the signals in our phone. At the same time, it is possible to configure the app, so that it sends the information acquired to our web service. In this way, the acquired information can be stored in a database for later analysis. At the same time, our web service uses these data, to enhance the respective recommendations of activities.

Once the signals have been acquired and pre-processed, they are sent to the web-service. This web-service uses different AI tools to analyse the signals and to try to detect emotional states, stress, or heart problems such as arrhythmia, tachycardia or bradycardia.

2.2 Software Description

To suggest exercises, it is used the FitCLA [13,19,20], a cognitive assistant that profiles the users and adapts the exercises suggestion according to their cognitive and physical impairments (e.g. memory loss, assisted mobility). It prompts the users, using the wristband, with a suggestion at determined times. Furthermore, the caregivers can visualise the configurations and progression through a friendly website.

It uses the sensors available as a evaluation process, verifying if the user is performing the exercises within the expected parameters. Moreover, with the emotional information, obtained from the emotional detection module. This module use deep learning techniques to recognise the emotion [21]. The FitCLA is able to adjust classification of the exercises according to the emotional reactions of the users. Studies show that activities and exercises that involve cognitive and physical functions improve greatly their physical condition, memory ability and happiness levels [22,23].

The FitCLA is composed of the agenda manager (keeps the information of each user and manages the scheduling of information), the activities recommender (selects an activity/exercise from the database according to the algorithm's parameters), the module manager (the gateway for coupling new features) and the message transport system (communicating process of the different agents). The FitCLA tightly integrates with the hardware available (wristband, chest strap, and virtual assistant) using them to improve the suggestions and to communicate with the users.

3 Conclusions and Future Work

A cognitive assistant platform that aims to help people exercise has been presented in this paper. ME^{3}CA generates and adapts personalized exercises sequence for an individual, in this case, an elderly person at his/her home. To do this, the system incorporates a set of bio-sensors integrated in a chest strap and in a wristband. These sensors capture information that can be measured in the form of physical stress while a user is doing exercises, also, it is able to perceive the emotion of the person.

Currently, the approach is being tested in *Centro Social Irmandade de S. Torcato*, which is a daycare centre in the northern area of Portugal. For the time being, the robot is being used with a small number of patients and always under supervision of caregivers. As future work, we want to integrate new functionalities in order to detect whether the suggested exercises are completed by the user in a satisfactory way.

References

1. World population prospects: the 2017 revision, key findings and advance tables. Report, United Nations, Department of Economic and Social Affairs, Population Division (2018). https://esa.un.org/unpd/wpp/Publications/Files/WPP2017_KeyFindings.pdf
2. Conesa, J.C., Kehoe, T.J.: An introduction to the macroeconomics of aging. J. Econ. Ageing **11**, 1–5 (2018)
3. Ageing in the twenty-first century: a celebration and a challenge. Report, United Nations Population Fund (2012). https://www.unfpa.org/sites/default/files/pub-pdf/Ageing%20report.pdf
4. Ageing report: Europe needs to prepare for growing older. Technical report, European Commission: Department of Economic and Financial Affairs (2012)
5. Bettio, F., Verashchagina, A.: Long-term care for the elderly, provisions and providers in 33 European countries. Technical report, European Union (2013)
6. World alzheimer's report 2015: the global impact of dementia, an analysis of prevalence, incidence, cost and trends. Technical report, Alzheimer's Disease International (2015)

7. Smith, D., Lovell, J., Weller, C., Kennedy, B., Winbolt, M., Young, C., Ibrahim, J.: A systematic review of medication non-adherence in persons with dementia or cognitive impairment. PLOS ONE **12**(2), e0170651 (2017)
8. Leichsenring, K., Ilinca, S., Rodrigues, R.: From care in homes to care at home: European experiences with (de)institutionalisation in long-term care. Technical report, European Centre for Social Welfare Policy and Research (2015)
9. Kim, S.: Cognitive rehabilitation for elderly people with early-stage alzheimer's disease. J. Phys. Ther. Sci. **27**(2), 543–546 (2015)
10. Martinez-Martin, E., del Pobil, A.P.: Personal robot assistants for elderly care: an overview. In: Intelligent Systems Reference Library, pp. 77–91. Springer, Cham (2017). https://doi.org/10.1007/978-3-319-62530-0_5
11. Blue Frog Robotics. Buddy (2018). https://buddytherobot.com. Accessed 22 Oct 2018
12. InTouch Technologies. Intouch health (2018). https://www.intouchhealth.com/. Accessed 22 Oct 2018
13. Costa, A., Martinez-Martin, E., Cazorla, M., Julian, V.: PHAROS—PHysical assistant RObot system. Sensors **18**(8), 2633 (2018). https://doi.org/10.3390/s18082633
14. Chesta, C., Corcella, L., Kroll, S., Manca, M., Nuss, J., Paternò, F., Santoro, C.: Enabling personalisation of remote elderly assistant applications. In: Proceedings of the 12th Biannual Conference on Italian SIGCHI Chapter - CHItaly 2017. ACM Press (2017)
15. Carstensen, L.L., Turan, B., Scheibe, S., Ram, N., Ersner-Hershfield, H., Samanez-Larkin, G.R., Brooks, K.P., Nesselroade, J.R.: Emotional experience improves with age: evidence based on over 10 years of experience sampling. Psychol. Aging **26**(1), 21–33 (2011)
16. Wang, H., Wang, N., Yeung, D.-Y.: Collaborative deep learning for recommender systems. In: Proceedings of the 21st ACM SIGKDD International Conference on Knowledge Discovery and Data Mining, pp. 1235–1244. ACM (2015)
17. Christakou, C., Vrettos, S., Stafylopatis, A.: A hybrid movie recommender system based on neural networks. Int. J. Artif. Intell. Tools **16**(05), 771–792 (2007)
18. Salsekar, B., Wadhwani, A.K.: Filtering of ECG signal using butterworth filter and its feature extraction. Int. J. Eng. Sci. Technol. **4** (2012)
19. Costa, A., Rincon, J.A., Carrascosa, C., Julian, V., Novais, P.: Emotions detection on an ambient intelligent system using wearable devices. Futur. Gener. Comput. Syst. (2018)
20. Rincon, J.A., Costa, A., Villarrubia, G., Julian, V., Carrascosa, C.: Introducing dynamism in emotional agent societies. Neurocomputing **272**, 27–39 (2018)
21. Rincon, J.A., Costa, A., Novais, P., Julian, V., Carrascosa, C.: Intelligent wristbands for the automatic detection of emotional states for the elderly. In: Intelligent Data Engineering and Automated Learning, IDEAL 2018, pp. 520–530. Springer, Cham (2018)
22. Nacke, L.E., Nacke, A., Lindley, C.A.: Brain training for silver gamers: effects of age and game form on effectiveness, efficiency, self-assessment, and gameplay experience. CyberPsychology Behav. **12**(5), 493–499 (2009)
23. Ertel, K.A., Maria Glymour, M., Berkman, L.F.: Effects of social integration on preserving memory function in a nationally representative US elderly population. Am. J. Public Health **98**(7), 1215–1220 (2008)

A New Conductivity Sensor for Monitoring the Fertigation in Smart Irrigation Systems

Javier Rocher[1], Daniel A. Basterrechea[1], Lorena Parra[1,2], and Jaime Lloret[1(✉)]

[1] Instituto de Investigación para la Gestión Integrada de zonas Costeras, Universitat Politècnica de València, C/Paraninf, 1, 46730 Grao de Gandia, Valencia, Spain
{jarocmo,loparbo}@doctor.upv.es, dabasche@epsg.upv.es, jlloret@dcom.upv.es
[2] IMIDRA. Finca "El Encin", A-2, km 38,2, 28800 Alcalá de Henares, Madrid, Spain

Abstract. The incorrect fertilization of the crops can cause problems in the environment and extra costs. A solution is to perform fertigation controlling the amount of fertilizer in the water. In this paper, we test different combinations of coils for determining the amount of fertilizer in the water. A coil is powered by a sine wave of 3.3 peak-to-peak Volts for inducing another coil. These sensors will be included in a smart irrigation tube as a part of a smart irrigation system based on the Internet of Things (IoT). The aim of this system is to detect different sorts of problems that can cause incorrect fertilization, which affects the sustainability of agriculture. This system can be used in different scenarios where tubes are used to irrigate. We present the performed test to evaluate the suitability of the created prototypes. At first, we test with different dilutions of NaCl (table salt) and, after it, we performed tests with nitromagnesium (a fertilizer). We checked that at the same salinity the induction value changes if it is found in water with NaCl or nitromagnesium. Of all the tested prototypes it is concluded that the prototype P2 is the most optimal g/L because there is a difference in the induced voltage between 0 and 45 g/L of nitromagnesium of 3.79 V with a good correlation coefficient. In addition, the average error in the different samples tested in the verification test is 2.15%.

Keywords: Coils · Conductivity · Fertigation · Irrigation systems · IoT

1 Introduction

Agriculture provides most of the food consumed worldwide today. It is estimated that 95% of the food consumed is produced directly or indirectly in the soil. A healthy soil for optimal growth of species must have a certain amount of microorganisms that transform inert matter, such as minerals into nutrients. To maintain an optimal relationship in the soil it must have an optimal ratio of C: N (carbon/nitrogen), so that

P. Novais et al. (Eds.): ISAmI 2019, AISC 1006, pp. 136–144, 2020.
https://doi.org/10.1007/978-3-030-24097-4_17

existing bacteria can provide nutrients to the crops, these two compounds being limiting in the soil [1].

Intensive agriculture is one of the main causes of soil degradation. This type of soil loses most of the nutrients, because of this intensive use of lands. Consequently, future plantations lack this support to produce optimal crops [2]. Due to this need for plants to maintain a correct balance of nutrients in the soil, the use of nitrogen fertilizers has increased to a large extent among farmers.

The use of these compounds has some risks, caused by undue fertilization, which leads to serious problems both in the quality of soil, by the loss of nitrogen-fixing microorganisms [3], and in the environment, such as contamination of the soil, groundwater or the pollution of rivers by runoff [4]. Groundwater represents a source of fresh water of great importance for human beings, due to its use for supplying the population [5], and its participation in hydrological processes. Over-fertilization causes insoluble nitrogen to percolate through the soil in the form of nitrate (NO_3) through leaching, infiltrating to the root zone, and reaching aquifers which would be contaminated [6]. Thus, it is crucial to develop sensors to monitor the amount of fertilizer in order to have a smart fertigation system. The IoT system, composed of different sensors placed in each three of the plots, was already described in [7]. In our system, we have different sort of sensors, including soil moisture, soil temperature, and soil humidity sensors. In addition, some sensors are included in the tube itself as the sensors presented in this paper. All the sensors will be wirelessly connected to a central Access Point (AP). The AP will receive the information and will forward it to an upper layer of the architecture in order to implement artificial intelligence techniques to define different rules to enhance its efficiency.

In this paper, we propose a conductivity sensor, based on a magnetic coil, which is capable of differentiating with great precision the changes of conductivity in a small range of salt concentration. Our proposal is ideal for monitoring the flow of fertilizers in irrigation water in a precise way using changes produced in the electromagnetic field produced by the developed coil. If the amount of fertilizer increases in the irrigation water, the field produced by the sensor will be modified, bringing abrupt changes in the conductivity and varying the mS/cm. This is a very simple, but, at the same time, efficient way to obtain real-time data of what happens in the water destined for the fields and finally to control the over-fertilization of the fields, reducing the options to contaminate the environment by the N of the soil.

The rest of the paper is structured as follows. Section 2 presents some published related works. Section 3 presents the architecture and sensors of our system. The results of our system are presented in Sect. 4. Finally, Sect. 5 details the conclusion and future work.

2 Related Work

This section presents different related work focused on the water quality monitoring and the main drawbacks which cause that those solutions cannot be applied to our case.

Ramos et al. [8] developed and tested an in-situ conductivity sensor using cells for water quality monitoring. The sensor consists of three electrodes and two terminal devices. This system works by exchanging ions creating an electric field. When the water flow passes through this system, the electric field varies. This sensor works correctly although it depends largely on an anti-fouling coating, which despite being a very small amount. Also, this device is not suitable for continuous measurements in a closed space such as an irrigation pipe, maintaining a high precision in low salinity ranges.

Parra et al. [9] developed a conductivity sensor based on coils. A part of this is fed with a current at a specific frequency and voltage, creating a magnetic flux that induces the other part of the coil, thus creating a magnetic field. This field varies according to the presence of salts in the water and thus can be measured. This sensor has been tested in water with different concentrations of NaCl being a device with an optimal response for use in marine spaces. On the other hand, no experimental tests have been carried out with fertilizer and irrigation water, which is why this development is not suitable for use in agriculture.

Gong et al. [10] proposed a conductivity sensor based on two flat electrodes built on a PBC board. The author makes the measurements using KCI and water MQ. This sensor works at 334 kHz obtaining significant variations in a wide range of salinity, suitable for marine environments. Although the device offers good results, it has not been tested in irrigation water with fertilizer, which represents completely different conditions to the one carried out in the experiment. Also, the cost presented is greater than what is wanted by a sensor of these characteristics, since it operates at a higher frequency than that used by our sensor.

Parra et al. [11] proposes the development of a low-cost wireless conductivity sensor to control freshwater flooding of lakes and mangrove reserves. The sensor used is composed of a toroidal coil and a solenoid, so that the first is fed by a sinusoidal wave and this induces the second. The results show a wide range of detection able to detect differences between small and large amounts of salt, typical of the sea. However, the need to detect fertilizer in the water requires a higher precision of the sensor in low concentrations, therefore this would not be optimal for this purpose.

There are many solutions to monitor conductivity in water. Nonetheless, none of them has tested their sensors with fertilizers. In addition, we need a sensor which can measure a low range of conductivities, to distinguish variations of fertilizers in water.

3 Proposal

In this section, we are going to describe the proposed prototypes to measure the concentration of fertilized and the proposed architecture of our IoT system.

Firstly, prototypes are described. They are shown in Table 1; all of them are created with copper wire with 0.4 mm of diameter. The prototypes are coiled over a PVC tube with 25 mm of diameter and 3 mm of thickness. The differences between them are mainly the number of spires in the powered coil (PC) and induced coil (IC) and the distribution of the turns in different layers. Those characteristics can be seen in Table 1. The employed coils are based on the best results obtained by Parra et al. [12].

Our system is composed of different sort of sensors which are responsible for monitoring several parameters from the soil, the water, and the fertilizer. We have sensors monitoring the soil, which measures the moisture, the temperature, and the conductivity. The conductivity of the soil is related to the water quality used for irrigating and to the utilized fertilized. In addition, the quality of the water before adding the fertilizer is monitored considering the following parameters: temperature, conductivity, and turbidity. An Arduino sensor is selected to gather the data from the sensors, and it sends the information to an AP using WiFi technology. Once the data is received, the AP forwards it to a Database where AI is applied to determine the best way to fertigate the plants to enhance the sustainability of the fertigation process. This architecture and these sensors can be used in any sort of agriculture, which uses tubes to irrigate since the drip irrigation in orange trees to sprinklers in urban lawns.

Table 1. Characteristics of the prototypes.

Name	Picture	Coils	Name	Picture	Coils
P1		Spires: 40 PC, 80 IC Layers: 2	P5		Spires: 40 PC, 100 IC Layers: 1
P2		Spires: 40 PC, 80 IC Layers: 4	P6		Spires: 40 PC, 80 IC Layers: 1
P3		Spires: 40 PC, 80 IC Layers: 8	P7		Spires: 40 PC, 60 IC Layers: 1
P4		Spires: 80 PC, 160 IC Layers: 2	P8		Spires: 40 PC, 40 IC Layers: 1

4 Results

In this section, we show the results of different coils in NaCl. Then we test with a fertilizer (Nitromagnesium 22(5)). Finally, we select the best prototype.

4.1 Tests with Different Prototypes

In this subsection, we show the tests with different prototypes with five samples which have different concentrations of table salt, also known as sodium chloride (NaCl). The objective of this test is to find the working frequency (WF), it is to say, the frequency which has the maximum difference between samples of the prototypes.

For test the different prototypes we prepared 5 samples with a concentration of 0, 5, 10, 20, 35 and 45 g/L of NaCl. The conductivity of these samples is 0.37, 9.28, 16.21, 28.7, 48.5, and 60.3 mS/cm respectively (measured with an EC meter model Basic 30). The powered coil is connected to a wave generator model AFG1022 and the voltage of the IC is measured with an oscilloscope model TBS1104. The wave generator generates a sine wave of 3.3 peak-to-peak volts. Measurements are performed at frequencies from 10 to 1000 kHz each 10 kHz. In the PC, there is a resistance of 47 ohms in the positive cable placed in series. In the IC a capacitor of 10 nF in parallel is utilized.

Table 2. WF and IV variations for the models tested with NaCl

Prototype	WF (kHz)	IV variation (V)	Prototype	WF (kHz)	IV variation (V)
P1	140	3.67	P5	140	2.19
P2	110	−3.96	P6	160	1.56
P3	90	−6.58	P7	180	−5.36
P4	270	−1.76	P8	260	3.60

The induced voltage (IV) in all the prototypes changes when the conductivity varies in all the frequencies tested. Therefore, it will be necessary to evaluate which will be the best prototype and the WF. To determine the WF of one prototype we must accomplish the following characteristics, (I) large difference in IV; (II) values can be adapted to a mathematical model; and (III) the IV follows the same trend in all samples. Table 2 shows the WF and the difference of IV between 0 and 45 g/L of NaCl.

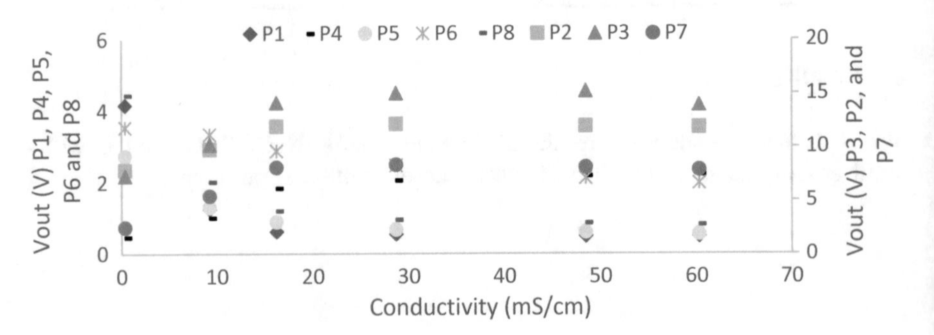

Fig. 1. IV of the different prototypes in the best frequency with NaCl.

In Fig. 1 we can observe the values of IV for the different prototypes in their working frequency. All prototypes have a similar trend. They present a big change between the values 0 to 16.21 mS/cm. From this point, the IV reaches an upper limit and there are no more increments in the IV. In the prototypes P3 and P7, we can be observed a change in the trend compared to the other prototypes. Moreover, in prototypes P4 and P6, the difference between the dissimilar samples is small if we compared with the other prototypes. This causes a decrease in sensor sensitivity.

In this case, the best prototype is the P3. It has a large difference in the IV between the lowest and highest salinity and a good R^2 (0.9113) (R^2 is a statistic expression that indicates how the mathematical model adapted to the different points is). The mathematical equation that models the IV of prototype P3 is presented in Eq. 1.

$$Vout(V) = -0.003 \times Cond.(mS/cm) + 0.2352\ Cond.(mS/cm) + 8.0063 \quad (1)$$

4.2 Calibration of Selected Prototypes

In this section, we show the effect in the coils when the conductivity changes.

In this case, we prepared 5 samples with different amounts of fertilizer, Nitromagnesium 22(5) The used nitromagnesium has 22% nitrogen and 5% magnesium. Finding 11% of nitrogen in nitric form and the other 11% shaped like ammoniac. These samples are prepared with a concentration of 0, 5, 10, 20, 35, and 45 g/L of fertilizer. The conductivity of these samples is 0.37, 5.28, 10.03, 17.84, 28.6, and 36.5 mS/cm. Also, the values of conductivity are less than in the case of NaCl for two reasons. We use commercial nitromagnesium that contains a quantity of non-soluble matter so that the farmer can manipulate it. In addition, the electrical conductivity of each substance is different. The values of IV in the different prototypes, in the same frequencies than we used before can be seen in Fig. 2. The coils behave differently if they are in the water with NaCl or with nitromagnesium. The coils work differently depending on the diluted salt. Prototypes P5 and P7 presented and a low R^2 (0.6027 and 0.7299 respectively), for this reason, these models are not good for the monitoring of fertigation. In addition,

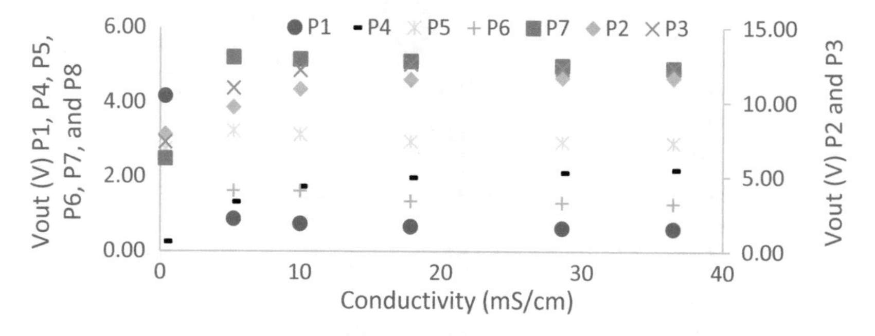

Fig. 2. IV of the different prototypes in the best frequency with nitromagnesium.

prototype P5 has a low voltage difference in the diverse samples. So, this prototype is not useful for our purpose. Prototypes P6, P7, and P8 have a difference of 1.80, −2.43, 2.73 V. While prototypes P1, P2 and P3 have differences of 3.56, −3.79, −4.95 V. These 6 prototypes adapt well to a mathematical model with high R^2 coefficients and have a good induced voltage difference between the samples. Therefore, we select prototypes P1, P2, and P3 for its verification. We select these prototypes because are the prototypes that present more difference between the samples 0 and 45 g/L of nitromagnesium. So, prototypes P6, P7 and P8 will have greater precision.

Equations 2, 3 and 4 is the mathematical model of P1, P2, and P3. These have an R^2 of 0.9417, 0.9735, and 0.9223 respectively.

$$Vout(V) = 2.1684 \times cond.(mS/cm) - 0.405 \tag{2}$$

$$Vout(V) = 8.5715 \times cond.^{.0915}(mS/cm) \tag{3}$$

$$Vout(V) = 1.3266 \times ln(cond.\ (mS/cm)) + 8.2649 \tag{4}$$

4.3 Verification Test

Finally, we are going to analyze the precision of the selected prototypes. For this test, we prepared different samples from 1 to 30 g/L with values from 1.99 to 24.6 mS/cm.

In Table 3, the values of IV, the values of the mathematical model, and the difference between these values are shown from different prototypes. The error is the difference between the real value and the value of the model. The relative error is the error divided for real value and multiplied by 100. The prototype P2 has a lower relative error with an average error of 2.15%. For this reason, we selected this prototype as the best one.

Table 3. Verification of prototypes P1, P2 and P3

Conductivity (mS/cm)	Difference (V)			Difference (%)		
	P1	P2	P3	P1	P2	P3
1.99	0.15	0.04	0.80	10.38	0.39	9.61
3.05	0.30	0.19	1.62	28.21	2.07	14.27
4.58	0.12	0.12	0.18	11.51	1.22	1.75
6.95	0.17	0.06	0.70	20.79	0.54	6.04
11.59	0.09	0.34	0.72	12.06	3.08	5.87
13.56	0.05	0.49	0.88	7.56	4.27	6.96
16.09	0.03	0.45	0.62	4.74	3.89	4.90
19.2	0.00	0.27	0.35	0.12	2.33	2.78
24.6	0.19	0.19	0.55	24.40	1.68	4.23
Average	0.11	0.24	0.64	13.06	2.15	5.62

5 Conclusion

Fertigation is a solution that allows irrigation to be managed in order to reduce the environmental problems of farming systems. In this article, we presented a new use of an existing sensor based on coils for fertigation monitoring.

We have determined the working frequency of the different prototypes with common salt. Our results indicate that the best prototype was the P3. Subsequently, we have tested the same with nitromagnesium, and the gathered data show different values than the data with the NaCl. It indicates that the coils can be more sensitive to different types of salts. In this case, the best models to work with the fertilizer were P1, P2, and P3. The P2 is the prototype with the smallest relative error during the verification test. In future work, we want to improve the sensitivity of our sensor and different salts.

Acknowledgment. This work has been partially supported by the European Union through the ERANETMED (Euromediterranean Cooperation through ERANET joint activities and beyond) project ERANETMED3-227 SMARTWATIR by the "Ministerio de Educación, Cultura y Deporte", through the "Ayudas para contratacion predoctoral de Formación del Profesorado Universitario FPU (Convocatoria 2016)". Grant number FPU16/05540 and by the "Fondo Europeo Agrícola de Desarrollo Rural (FEADER) – Europa invierte en zonas rurales", the MAPAMA, and Comunidad de Madrid with the IMIDRA, under the mark of the PDR-CM 2014-2020 project number PDR18-XEROCESPED.

References

1. Simons, J., Dannenman, M., Pena, R., Gessler, A., Renmenberg, H.: Nitrogen nutrition of beech forests in changing climate: importance of plant-soil-microbe water, carbon, and nitrogen interactions. Plant Soil **418**(1–2), 89–114 (2017)
2. Tsiafouli, M., et al.: Intensive agriculture reduces soil biodeversity across Europe. Glob. Chang. Biol. **21**(2), 973–985 (2015)
3. Wood, S.A., Almaraz, M., Bradford, M.A., McGuire, K.L., Naeem, S., Neill, C., Palm, C.A., Tully, K.L., Zhou, J.: Farm management, not soil microbial diversity, controls nutrient loss from smallholder tropica agriculture. Front. Microbiol. **6**, 90 (2015)
4. Lentz, R., Carter, D., Haye, S.: Changes in groundwater quality and agriculture in forty years on the Twin Falls irrigation tract in southern Idaho. Soil Water Conserv. **73**(2), 107–119 (2018)
5. Fernández, R., Fernández, J.A., López, B., López, J.A.: Aguas subterráneas y abastecimiento urbano, 1st edn. IGME, Castilla-La Mancha (2000)
6. Lawniczak, A.E., Zbierska, J., Nowak, B., Achtenberg, K., Grzeskowiak, A., Kanas, K.: Impact of agriculture and land use on nitrate contamination in groundwater and running waters in Central-West Poland. Environ. Monit. Assess **188**, 172 (2016)
7. García, L., Parra, L., Jimenez, J.M., Lloret, J., Lorenz, P.: Practical design of a WSN to monitor the crop and its irrigation system. Netw. Protoc. Algorithms **10**(4), 35–52 (2018)
8. Ramos, H., Gurriana, L., Postolache, O., Pereira, M., Girão, P.: Development and characterization of a conductivity cell for water quality monitoring. In: IEEE 3rd International Conference on Systems, Signals and Devices (SSD), Sousse (2005)

9. Parra, L., Sendra, S., Ortuño, V., Lloret, J.: Water conductivity measurements based on electromagnetic field. In: 1st International Conference on Computational Science and Engineering (CSE 2013), Valencia, pp. 139–144 (2013)
10. Gong, W., Mowlem, M., Kraft, M., Morgan, H.: Oceanographic sensor for in-situ temperature and conductivity monitoring. In: Oceans 2008-Mts/IEEE Kobe Techno-Ocean (OTO 2008), Kobe, pp. 1–6 (2008)
11. Parra, L., Sendra, S., Lloret, J., Rodriguez, J.: Low cost wireless sensor network for salinity monitoring in mangrove forests. In: SENSORS 2014, Valencia, pp. 126–129 (2014)
12. Parra, L., Sendra, S., Lloret, J., Bosch, I.: Development of a conductivity sensor for monitoring groundwater resources to optimize water management in smart city environments. Sensors **15**(9), 20990–21015 (2015)

Dynamic Rules Extraction in Big Data Context for Knowledge Capitalization Systems

Badr Hirchoua[1(✉)], Brahim Ouhbi[1], and Bouchra Frikh[2]

[1] National Higher School of Arts and Crafts (ENSAM), Industrial Engineering and Productivity Department, Moulay Ismail University (UMI), Meknes, Morocco
hirchoua.badr@gmail.com, ouhbib@yahoo.co.uk

[2] Higher School of Technology (EST), Computer Science Department, Sidi Mohamed Ben Abdellah University (USMBA), Fez, Morocco
bfrikh@yahoo.com

Abstract. Big data is particularly challenging, when focusing on pattern mining to find rules that describe the hidden behavior over time. Traditional formalism for rules extraction paradigms, has been extended to a high abstraction level, and improved with the automatic choice of the feature space dimension. This paper presents a novel comprehensive theory of large-scale learning with β random walk, and variational autoencoder. The new theory has the following components: 1. Rethinking learning theory; it validates the two bounds context, the local and the global by which the knowledge behavior is caught. 2. Hidden features extraction; large scale variational autoencoder provides a complete decentralization of the latent distribution resided in the latent space. Thus, as a result a new representation of the high dimensionality is replaced by a more relevant low dimensionality distribution. 3. Rules construction, the optimal bound of pattern recognition is achieved by a high abstraction level. In that sense, the proposed theory provides a new understating of the benefit of the hidden features, and gives concrete response to the diversity of rules in the big data context. The results show that the extracted rules are solid by achieving high accuracy, as well as, a high precision.

Keywords: Knowledge capitalization · Rules extraction · Graph theory · Knowledge sharing · Variational autoencoder · Random walk

1 Introduction

In the age of big data, the rules are changing over a small times slot. For example, rules in quantitative finance, Facebook friends' graph are constantly evolving. Features that are predictive in some time-period might lose their edges, while

P. Novais et al. (Eds.): ISAmI 2019, AISC 1006, pp. 145–152, 2020.
https://doi.org/10.1007/978-3-030-24097-4_18

other patterns emerge. Unsatisfying way of modeling, as there are certain shortfalls, is that the estimation should be long so as to incorporate as much training data as possible. Besides, the estimation should be short to incorporate only the most recent data. The extracted rules help to understand insights in data, and catch the high-level semantics, this process is not a trivial task over big data. Rules are the summary of what a given data provides as knowledge. Big data context presents a different level of rules meaning. In other words, it contains several types of rules, so the extracted rules for a given type leads to ignore the other types, which are important. The key is to move to another level of abstraction in all extracted rules. The knowledge behavior is different from one knowledge base to another especially in big data.

This paper cuts with all these issues by proposing a fully new hybrid approach, which deals with the knowledge graph streams as a source of knowledge, and extracts a high-level abstraction rules, with the aim of catching the whole knowledge behavior. In terms of quality and uncertainty, the proposed model allows the training on much larger data sets ranging back a long time. One of the main challenges for rules extraction, at which knowledge capitalization is linked to, in addition to big data, is bringing deep construction to higher level abstraction. Learning deep rules representations, and deep generative models, follows the ambitious objective of disentangling the underlying causal factors. Furthermore, it explains the observed knowledge in different senses including the behavior and context construction.

In order to efficiently discover such high-level representations of abstract rules, the learning process is divided into two deep steps. The first step discovers the local context (node's neighbors), in jointly learning policies and representations, such that each of these policies can independently control the global context (nodes in the whole graph). Once the assumption of nodes weighting is finished, the second prior of hidden features selection on the high dimensions opens up. Features selection from latent variables is based on the hypothesis that thoughts are low-dimensional objects with a strong representation and explanatory power. A conscious thought thus selects a few abstract features using the variational autoencoder (VAE [4]), and combines them to make a stronger low dimensional space. Hidden features are more representative, especially when they are extracted from a series of knowledge behavior, as well as they impact the local, and the global context. Rules construction use these features to extract the accurate, and precise rules. Specifically the decision tree [9] algorithm is used to build a model from which the rules are extracted in an understandable and significant form.

The goal of this work is to learn a high-level rules by extracting the hidden features represented in continuous time space, without losing the knowledge behavior. More formally, given a knowledge graph $KG = (V_i, E_i)$, where V_i and E_i are the set of vertices and edges respectively, and $|V_i| = l$. Informally, KG is a collection of graphs $G_1, G_2, \ldots, G_n$. For all graphs, the vertices set $\cap_{i=0}^{l} V_i$, could have non-null intersection, to learn rule of each node across multiple graphs. The aim is to construct the rules using the recognized relevant patterns.

The novel approach proceeds with three major steps. First, the proposed ranking approach (Eq. 1) weights the graph nodes. The weighting operation requires meaningful neighbors on graphs, where similar vertices are gathered. While the weighing mechanism goes on, this is equivalent to a multi-scale graph clustering, that preserves local geometric structures, and weight each node in its local, and global context. This step produces a vector representation for each node. Second, the system fits the extracted vectors into a VAE neural network in order to extract the hidden features, as well as reduces the vectors dimensionality. Finally, the rules are formulated using decision tree algorithm.

The remainder of this paper is organized as follows. In Sect. 2, the related works of this research are reviewed. In Sect. 3, the contribution concepts and properties of different methods parts are presented, and the effectiveness of using them in rules extraction is discussed. Section 4 includes the results of the novel approach, and the proposed approach is also validated through different metrics. Section 5, concludes this work, and presents the proposals for future work.

2 Related Work

The relationship between knowledge sharing, and rules extraction has been widely carried by different researchers. Mohammadi et al. [7] proposed a framework which utilizes a mixture of data type (labeled and unlabeled) to converge toward rule-based policies. Moreover, they adapt the generative VAE to produce the action rank for the reinforcement environment.

Diving on the concept of decision rules, Ondel et al. [8] present a diagnosis method based on a multidimensional function and pattern recognition (PR). The training phase, consists in determining the feature space (a pattern vector), the decision space (the clusters), and developing a decision rule that produces borders between classes. Moreover, the decision phase consists in predicting an unknown pattern with one of the already defined clusters, according to the decision rule, which is developed based on the k-nearest neighbors' rule.

Soru et al. [10] introduce a hornconcerto algorithm, which acquires Horn clauses in large graphs without a schema. Their propositions can mine high-quality rules in diver tasks. Moreover, the ranked rules can perform significant inference much faster with less resources than embedding-based methods, and approaches that uses Markov Logic Networks. In the same vein, Yap et al. [12] demonstrate the usefulness of the Laplacian likelihood function for rule extraction capability to elicit the if-then rules from the trained network. In addition to that, a pruning procedure is used in order to reduce the number of categories, which have insignificant contribution to the overall network output. Those remaining categories are used for extraction of a compact, and meaningful rules set.

Starting from some defined parameters, Ehimwenma et al. [3] attempt to estimate and predict the required number of rules of a non-regular ontology model, by proposing a system of polynomial equation. Moreover, they focus on the estimation, and prediction of rules in non-regular ontology trees. This polynomial system pushes the knowledge, and the system engineer forward by knowing in advance the number of inductive learning clauses for a given system. Contrary to the researchers works [1,2], which predict the required number of classified induction rules for agent classification learning in regular ontologies, using a system of algorithmic equations. In order to emerge pattern mining, Vico et al. [11] describe, and analyze an algorithm for Multi-Objective Evolutionary, which deals with the Extraction of Fuzzy Emerging Patterns. Moreover, their approach returns a simple rules description of the emerging phenomena, but with better description of discriminative behavior.

Overall, these works highlight the need for two major steps. Firstly, the nodes weighting, similarly to Ondel et al. [8] work, which needs a pattern vector as a feature space. Secondly, the hidden features extraction, instead of reducing the number of categories as Yap et al. [12], it is better to represent the rules with a low dimensional features. Interestingly, the proposed approach uses these stages in order to construct an accurate rules, which signified the concept's relevance to the knowledge evolution over time.

3 The Proposed Approach

First and foremost, the powerful key in the proposed approach is the self-resource-based approach. In fact, the system only uses the information exiting in the knowledge graph, where every node has a behavior reflected locally by its neighbors, and globally by its weight over the whole graph. In big data context, there is a great need to look smartly how the local context affects the global context.

Every node serves a local context to its neighbors, and as a global context for the rest of the graph nodes. The main targeted goal is to converge to the stationary state, by which every node offers the hidden features, and context for its neighbors. The β random walk [6] proposed in this paper, ranks the sampled the node sequences using Eq. 1. However, the β random walk only considers the first-order proximity between nodes for each click, then uses this new weight to rank the neighbors clicks, and so on.

The start weight for every node at time $t = 0$ is: $\Psi_i(t) = \frac{1}{N_i}$, where N_i is the number of neighbors for the node i. Then, every node weight at the time $t = k/k = 1, \ldots$ is updated using the following proposed formula:

$$\Psi_i(t+1) = \Psi_i(t) + \sum_{j=0}^{N_i} (\beta(\Psi_j(t) - \Psi_i(t))) \tag{1}$$

Where $\Psi_j(t)$ is the neighbor node j weight, at time t, and β is the control hyper-parameter. The β is a hyper-parameter, which balance the latent paths selected by the traditional random walk. In other words, it controls the random walk, in order to learn an unseen paths in the graph. The algorithm chooses a different path once the β value is decreased, until the $\beta = 0$. To the end, the algorithm converges to a stationary distribution by which: $\Psi_i(t+1) = \Psi_i(t)$, or $\Psi_i(t+1) \simeq \Psi_i(t)$ in the same case where the $\beta \sim 0$. From another angle, $\forall \epsilon > 0, \exists t \in \mathbb{T}, \forall i: \ |\Psi_i(t+1) - \Psi_i(t)\ | < \epsilon$, where $\mathbb{T}$ is the time space. The small change can be noticed when the initial node weight is very small. As the initial weight gets bigger, the same small change cannot be noticed, the small noticeable change needed depends to the size of the weight.

The β hyper-parameter does not depend on time t as a continues variable space, but instead it controls the time by minimizing the walk space, and forcing the agent to take the possible paths from the minimized space. Particularly, local behavior approximations focus on sparse approximations comprising the local knowledge, which modify the global behavior each time it rectifies its weight. The global context approximations, which retain exact inference from the local approximations, and perform approximate inference controlled by a hyperparameter β. Liu et al. [5] devote the global approximations, which distillate the entire data (global context), and local approximations, which divide the data for subspace learning (local context).

Since the resulted vectors from the β random walk are presented in high dimensionality, the extraction of useful, important, and latent features is not a trivial task. Intuitively, the high dimensionality captures the semantics and handle the context, which helps to decide which dimensions are more representative before the extraction phase. This kind of features is formally called a latent variable, since the model does not necessarily know which settings are generating the best feature. The variational autoencoder which consists of two sub-networks named 'Encoder' and 'Decoder', is used to tackle the dimensionality reduction. Instead of mapping any input to a fixed vector, the VAE transforms it to a distribution, which accords to the future entries the same continues latent space to keep the exact extraction process.

The Fig. 1 represents the system architecture, which consists of three components, the node weighting, dimensionality reduction, and rules construction. The first step weights every node from the input knowledge graph, and outputs a high dimensional vectors representation. The second component is the VAE, which transforms the nodes vectors into a lower dimensional representation, and extracts the hidden features using the continuous latent space. The final step seeks the rules extraction using the decision tree algorithm.

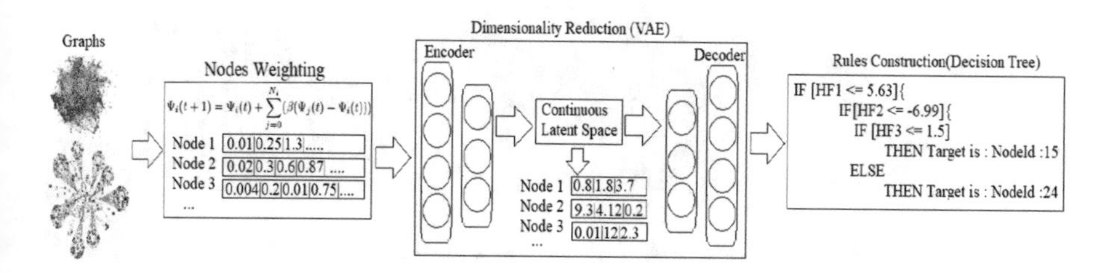

Fig. 1. The proposed system architecture.

4 Experimental Results

The graph used to evaluate the proposed approach, was generated using emails data collected for 18 months. Overall, it contains three million edge between three thousand different nodes. In the interest of evaluating the main contribution of nodes weighting, Fig. 2 presents the weighting mechanism for a set of three nodes $S_3 = \{w_i, w_j, w_k\}$, where the node w_i (grey color) is a local context to the node w_j (orange color), and the node w_j serves as a local context for the node w_k (blue color), as a result the node w_i is a global context for the node w_k. Clearly, the directly connected nodes (w_i, w_j) and (w_j, w_k) are following almost the same path. Moreover, each click (node in focus and its neighbors) in the graph presents a strong correlated distribution. Doubtless, the proposed β random walk for nodes weighting presents a strong solution to respond to different issues. Thus, Fig. 3 exemplify a click distribution using the compressed weighting over the whole process, it is understandably that the distribution is compact, and informationally dense.

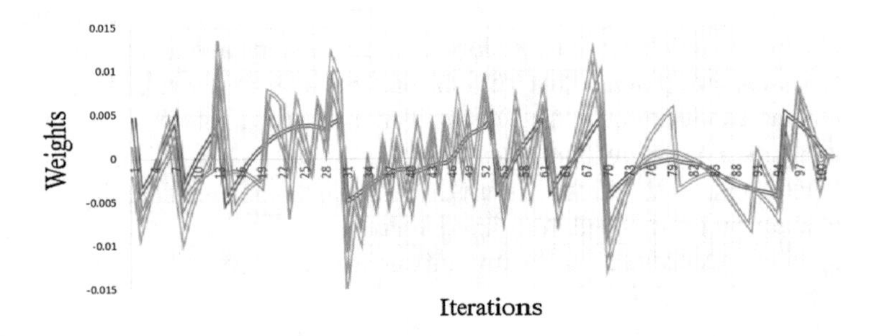

Fig. 2. Local and global context illustration.

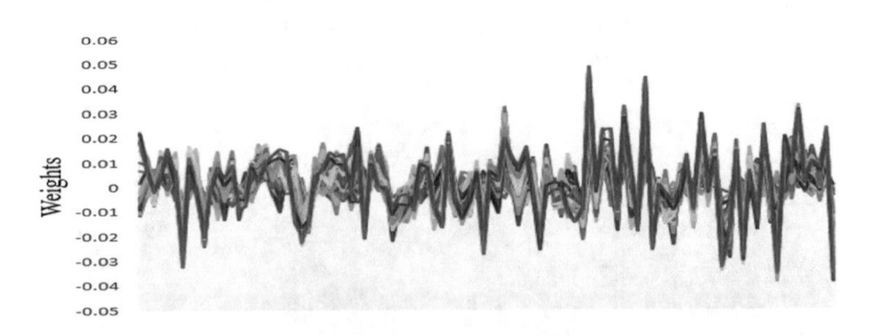

Fig. 3. Distribution of one click over a focus node.

The proposed system responds to the targeted requirements, by extracting the precise, as well as the understandable rules. First and foremost, the

Table 1. Model measures summary

Measure	Value	Formula
Precision	0.963	$Precision = \frac{TP}{TP+FP}$
Mean absolute error (MAE)	0.0014	$MAE = \frac{\sum_{i=1}^{n} \vert(P_i - O_i)\vert}{n}$
Root mean squared error (RMSE)	0.0317	$RMSE = \sqrt{\frac{\sum_{i=1}^{n}(P_i - O_i)^2}{n}}$

knowledge graph presents the correlation between different nodes, so the ranking mechanism must keep these correlations. The proposed β random walk maps the graphical representation to the vector representation. Therefore, this representation is fitted into the VAE to extract the hidden features as the new nodes representation. The resulted representation is reformulated in a final matrix, which contains the new representation for a given node as antecedent, and the consequence is the node which is connected to. Finally, the decision tree algorithm uses the resulted matrix to extract, and formulate the final rules. The final model achieves an accuracy of 90.17%, besides, the final results are presented in the Table 1, where:

- TP = true positives: number of examples predicted positive that are actually positive.
- FP = false positives: number of examples predicted positive that are actually negative.
- P_i is the predicted value, and O_i is the observed value.

An example of the extracted rule is presented in the following, where HF is the hidden feature, and NodeId represents the node index:

```
IF [HF1 <= 5.63]{
        IF[HF2 <= -6.99]{
                IF [HF3 <= 13.35]{
                        IF [HF1 <= 4.58] THEN Target is : NodeId :159
                        ELSE [HF1 > 4.58]THEN Target is : NodeId :167
                                        }
                ELSE{
                        IF [HF3 > 13.35] THEN Target is : NodeId :24
                        }
        }
        IF [HF2 > -6.99]{
                        IF [HF1 <= 4.58]  THEN Target is : NodeId :1
                        IF [HF1 > 4.58 ]THEN Target is : NodeId :4          }
}
```

5 Conclusion

The proposed approach restructures the input entities, which is composed of a large number of highly interconnected nodes, in more understandable rules.

Therefore, it combines the best of two worlds: the ability to slowly learn an abstract method for obtaining useful hidden features, via VAE, and the ability to extract the never-before-seen rules, over a new theory of contexts. The combination supports the extension of domains to which this process can effectively applied. The resulted accurate rules can capture many kinds of relations, as well as allow the ease knowledge sharing in big data context. This work is an end to end theory which can be applied to any problem in the range of knowledge capitalization context. Further research will seek the respond to speed up and handle the infinite knowledge graphs where the number of nodes goes to infinity.

References

1. Ehimwenma, K.E., Beer, M., Crowther, P.: Student modelling and classification rules learning for educational resource prediction in a multiagent system. In: 2015 7th Computer Science and Electronic Engineering Conference (CEEC), pp. 59–64. IEEE (2015)
2. Ehimwenma, K.E., Beer, M., Crowther, P.: Computational estimate visualisation and evaluation of agent classified rules learning system (2016). arXiv preprint: arXiv:1605.08878
3. Ehimwenma, K.E., Crowther, P., Beer, M.: A system of serial computation for classified rules prediction in non-regular ontology trees (2016). arXiv preprint: arXiv:1604.02323
4. Kingma, D.P., Welling, M.: Auto-encoding variational Bayes (2013). arXiv preprint: arXiv:1312.6114
5. Liu, H., Ong, Y.S., Shen, X., Cai, J.: When Gaussian process meets big data: a review of scalable GPS (2018). arXiv preprint: arXiv:1807.01065
6. Lovász, L., et al.: Random walks on graphs: a survey. Combinatorics, Paul Erdos is eighty, vol. 2(1), pp. 1–46 (1993)
7. Mohammadi, M., Al-Fuqaha, A.: Enabling cognitive smart cities using big data and machine learning: approaches and challenges. IEEE Commun. Mag. **56**(2), 94–101 (2018)
8. Ondel, O., Clerc, G., Boutleux, E., Blanco, E.: Fault detection and diagnosis in a set "inverter-induction machine" through multidimensional membership function and pattern recognition. IEEE Trans. Energy Convers. **24**(2), 431–441 (2009)
9. Quinlan, J.R.: Induction of decision trees. Mach. Learn. **1**(1), 81–106 (1986). https://doi.org/10.1023/A:1022643204877
10. Soru, T., Valdestilhas, A., Marx, E., Ngomo, A.C.N.: Beyond Markov logic: efficient mining of prediction rules in large graphs (2018). arXiv preprint: arXiv:1802.03638
11. Vico, A.M.G., Carmona, C.J., Gonzalez, P., Del Jesus, M.J.: MOEA-EFEP: multi-objective evolutionary algorithm for the extraction of fuzzy emerging patterns. IEEE Trans. Fuzzy Syst. **26**, 2861–2872 (2018)
12. Yap, K.S., Lim, C.P., Au, M.T.: Improved gart neural network model for pattern classification and rule extraction with application to power systems. IEEE Trans. Neural Netw. **22**(12), 2310–2323 (2011)

Real-Time Low-Cost Active and Assisted Living for the Elderly

António Henrique Almeida[1], Ivo Santos[1], Joel Rodrigues[1], Luis Frazão[1], José Ribeiro[1], Fernando Silva[1], and António Pereira[1,2](✉)

[1] School of Technology and Management, Computer Science and Communication Research Center, Polytechnic Institute of Leiria, 2411-901 Leiria, Portugal
{2160827,2160837,2090715}@my.ipleiria.pt,
{luis.frazao,jose.ribeiro,fernando.silva,
apereira}@ipleiria.pt

[2] Information and Communications Technologies Unit, INOV INESC Innovation, Delegation Office at Leiria, Leiria, Portugal

Abstract. The aging of population in recent years and the increase in life expectancy is raising challenges for finding new ways to guarantee healthy and controlled activities for the elderly. Most of them prefer living in their houses than in a community center, even if they live alone or isolated from their family; at home, their normal routine activities and comfort makes them feel well. In this paper, an Active and Assisted Living (AAL) solution to detect irregular situations in everyday life of the elderly living alone is presented. By using low-cost sensors in an Internet of Things (IoT) architecture we aim to gather data in specific areas of an elderly's house in order to give the system enough input to detect abnormal behavior. These sensors are non-intrusive to the elderly, do not disturb them, and do not force them to wear a device at all times. These sensors can also send information to edge computing devices that analyze the data in real time using machine learning algorithms and alert family or caretakers when an unusual situation arises. The proposed solution provides a system that monitors the main activities performed by the elderly and creates patterns based on that activity to achieve its results and is scalable in terms of sensors and data input.

Keywords: Active and Assisted Living · Edge computing · Internet of Things · Ambient Intelligence · Machine learning · Low cost · Real time

1 Introduction

Modern societies are facing a fast growth of aging population. Improved health and social care over recent years has increased life expectancy worldwide. Nearly 7% of the world's population is now over 65 years of age and the predictions indicate that older people will rise to approximately 20% by 2050 worldwide [1]. The number of elderly people living alone is increasing every day and typically they prefer to stay in their homes, which makes it even more difficult for family members to make sure that they

P. Novais et al. (Eds.): ISAmI 2019, AISC 1006, pp. 153–161, 2020.
https://doi.org/10.1007/978-3-030-24097-4_19

are okay. The concept of AAL is understood as the use of Information and Communication Technologies (ICT), a means of prolonging time, people can live in their preferred environment (usually their home), avoiding social isolation, giving them due support and promoting a better and healthier lifestyle. In this context, the concept of IoT can be used to support elders, particularly those who live alone in their homes.

The resulting data from IoT services can be processed by Machine Learning (ML) mechanisms that will extract knowledge from it in real time to create patterns about elderly's behavior. The proposed solution goals involve surveying the current state of the art related to the theme. To begin, an identification of low-cost and non-intrusive sensors, adequate to the proposed problem, is done. A survey of tools and mechanisms concerning the use Artificial Intelligence (AI)/ML to extract knowledge from IoT data in real time is also presented. A prototype of sensory nodes for acquisition and transport of information and subsequent storage in the Platform is built. Concluding, a module of real-time knowledge extraction of IoT data using a studied algorithm is applied.

In this work, we present a system that aims to apply state of the art concepts, in order to provide an AAL solution for elders, which includes low-cost equipment and offers real-time communication with caregivers. The specific contributions of our work include a system that is available anytime and anywhere for family and caregivers, uses low cost devices to be affordable by anyone, applies machine learning algorithms for the data being gathered by the IoT sensors which allow the system to detect unusual behavior, and finally, it is adapted for elders that live alone in their homes.

The rest of the paper is organized as follows. Section 2 presents the fundamental concepts and some works related to assistive home systems. The general architecture of the proposed solution is described in Sect. 3. In Sect. 4, the implementation of a functional scaled model of a smart home for the elderly is presented. In Sect. 5, Analysis and Discussion are presented. Finally, in Sect. 6, the conclusions are drawn and some ideas for future work are presented.

2 Background

The work presented by Rashidi et al. [2] states that, in recent years, we have witnessed a rapid surge in assisted living technologies due to a rapidly aging society. AAL encompasses technical systems to support elderly people in their daily routine to allow an independent and safe lifestyle for as long as possible. The main goal of AAL is to maintain and foster the autonomy of those people and, thus, to increase safety and quality in their lifestyle and in their home environment. AAL applications include services, products and concepts that allows achieving that goal.

The work presented by Remagnino et al. [3] indicates that **Ambient Intelligence** (AmI) was introduced by the European community ([1, 4]) to identify a paradigm to equip environments with advanced technology and computing to create an ergonomic space for the occupant user. An AmI system requires the use of distributed sensors and actuators to create a pervasive technological layer, able to interact transparently with a user, either passively by observing and trying to interpret what the user actions and intentions are, or actively by learning the preferences of the user and adapting the system's parameters to improve the quality of life and work of the occupant.

The work presented by Kotsiantis, Zaharakis and Pintelas et al. [5] says that there are several applications for ML, the most significant of which is data mining. People are often prone to making mistakes during analyses or, possibly, when trying to establish relationships between multiple features. This makes it difficult for them to find solutions to certain problems. ML can often be successfully applied to these problems, improving the efficiency of systems and the designs of machines. Every instance in a dataset used by ML algorithms is represented using the same set of features. The features may be continuous, categorical or binary. If instances are given with known labels (the corresponding correct outputs) then the learning is called supervised, in contrast to unsupervised learning, where instances are unlabeled. By applying these unsupervised (clustering) algorithms, researchers hope to discover unknown, but useful, classes of items. Another kind of ML is reinforcement learning. The training information provided to the learning system by the environment (external trainer) is in the form of a scalar reinforcement signal that constitutes a measure of how well the system operates. The learner is not told which actions to take, but rather must discover which actions yield the best reward, by trying each action in turn.

Considering the work presented by Gazis et al. [6], **IoT** is a socio-technical phenomenon with the power to disrupt our society such as the Internet did before. IoT promises the (inter-) connection of myriad of things providing services to humans and machines. In [7], IoT is described as a technology that can provide a large amount of data about human, objects, time and space. Recent IoT development on existent Internet technologies create large opportunities to innovative existing services based on low-cost sensors and wireless communication.

The work from Yu et al. [8] states that nowadays millions of sensors and devices are continuously producing data and exchanging important messages. **Edge computing** will approximate data computation and storage to the end users. That way, the nodes installed in a network can significantly reduce the latency in message exchange, relieving the computational stress aggregated to the data center (cloud). Furthermore, transferring the computation process from nodes that have limited resources to others that are more powerful, can extend system lifetime.

The work presented by Blackstock and Lea [9] describes **Node-RED** as a web-based tool for connecting hardware devices and Application Programming Interfaces (API). It uses the concept of flows that are constituted by nodes connected by wires. The user interface consists of a flow editor with node templates on the left that can be dragged and dropped into a flow canvas. All users accessing Node-RED manage a single flow, which may be shown on multiple pages.

According to Manandhar [10], **Message Queuing Telemetry Transport** (MQTT) is a lightweight protocol, data-centric and specially designed for resource constrained devices like IoT devices. All the devices publish and subscribe the messages to the central broker, and this broker will handle delivering messages to subscribers. Devices do not need to execute simultaneously, and they can publish and subscribe asynchronously. These principles also turn out to make the protocol ideal of the emerging IoT world of connected devices.

The works presented below apply the abovementioned concepts to the development of AAL solutions for the Elderly.

The work presented by Aran et al. [11] presents a framework to analyze elderly daily behavior using only motion and state-change sensors. As an almost seamless and unobtrusive setting, it provides a promising approach for adoption among elderly and has shown to sufficiently capture day to day activities in real settings. The analysis used on the study is based on data gathered from 40 elderly homes on a four-month period. In addition to the sensor data, there are annotations from questionnaires and daily activity journals.

The work presented by Demir et al. [12] specifies an IoT solution that contains multiple sensors distributed by the elder's home. It uses pressure sensors in the bed and in the kitchen's table chair to detect when the elder is sleeping or eating, respectively. It resorts to light sensors to alert the elder when a light is left on by mistake. A thermal sensor is placed in the bathroom to verify the time that the elder spends in that room. Provides a rain sensor in front of a window with the purpose of warning the elder when it is open and raining. Finally, the cloud service "pushing box" is used in order to store the information and notify the caregiver.

The review presented by Dawadi et al. [13] was designed with the purpose of helping IoT developers. It provides an overview of the existing work developed in this area, gathering a set of papers related to IoT for the elder.

3 The Proposed System

This section presents and describes the development and implementation of a prototype in order to demonstrate the communication between a sensor and the platform. Figure 1 shows the architecture of the implemented system.

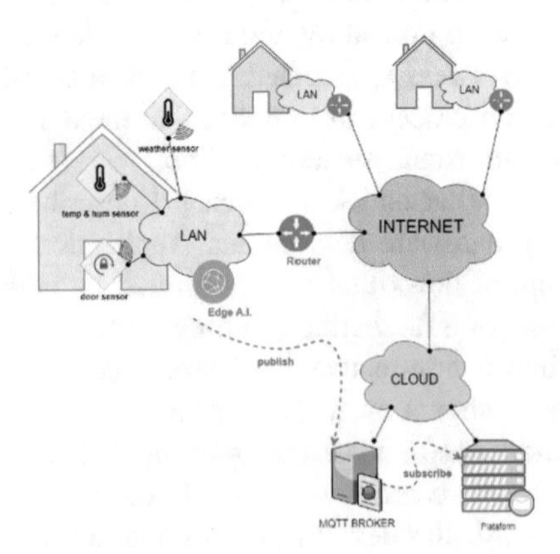

Fig. 1. Architecture of proposed system

The proposed system implements a solution where multiple houses can coexist. Each house has a local area network (LAN) and a set of non-intrusive sensors that will be responsible for measuring the most important variables of the quotidian. Sensors such as temperature meter, carbon dioxide meter, energy or water consumption meter, door opening detection and movement detection can be used to gather important information about elder habits. These sensors are connected to resource constrained IoT devices with communication capabilities that are programmed to control and read values from sensors. The IoT device sends information from the sensors through an 802.11 network to a single-board computer (SBC) on the same network (as shown in Fig. 1 as Edge A.I). This SBC device acts as an Edge Hub, which is an intermediary device responsible to implement edge computing solution. The parcel of data that was not discarded by the Edge Hub in the process of edge computing will be sent to a cloud-based platform to be used in a supervised AI algorithm. In order to achieve this connectivity between an Edge Hub and the cloud platform, all LANs connect to an Internet provider.

The communication between devices use the MQTT protocol, and the Edge Hub has the function of a MQTT broker. The broker is responsible for receiving and filtering the messages, determining who is subscribed to each message, and sending the message to these subscribed clients. Each LAN device sends MQTT publish messages with information about data gathered in its house sensors, while the cloud platform receives MQTT subscribe messages from all LAN devices. At the platform, a supervised mechanism is applied to all data received, in order to create patterns regarding the elder's behavior, detect unusual actions and warn family and caregivers.

In summary, the proposed system and architecture are based on edge computing. In this concept, some processing of the data is accomplished using one of the edges of the system. The relevant data is produced on devices modules at the edge of the network, making it more efficient, in terms of required system communication bandwidth, to process this data in the end devices rather than in cloud servers. However, this solution will also be using cloud computing, because the edge devices do not have the computational power to solve some of the more difficult tasks. Whenever an abnormal situation is detected the system activates its corresponding support services. This solution aims to help elders that live all by themselves, since it gives them a better quality of life with real comfort and security as well as to their relatives and caregivers.

4 Implementation and Testing

This section presents and describes the development and implementation of a functional prototype in order to demonstrate the technical and economic feasibility of the communication system. Figure 2 presents the diagram of the implemented prototype system. The elders are expected to present similar behavior in some common actions. Examples of such common actions include leaving the house at the same time every day to take care of their garden, buying bread, drinking their coffee at a local shop or visiting and feeding their livestock animals. In order to detect whenever the elder leaves the house a magnetic door sensor is used in order to detect when a door opens or closes. The sensor used for this purpose is a magnetic reed switch (see Fig. 3) which is installed at the door of the house. Each half of the sensor contains a magnet, so that when they are separated (doors opening), its value is different to when they are close together (door closed).

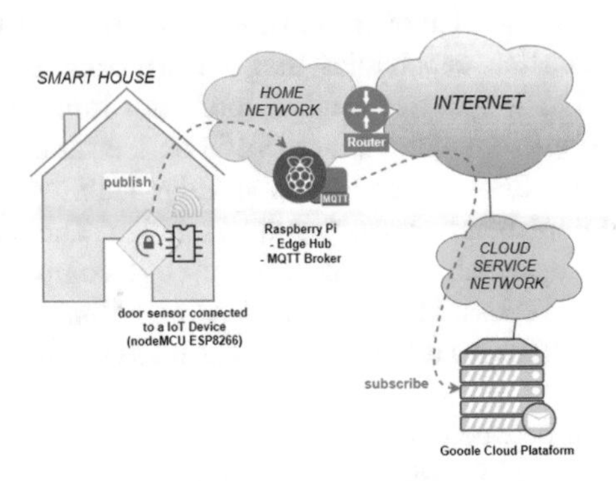

Fig. 2. Implemented prototype system

Other sensors are used to acquire information about the weather. In days of bad weather, the elder may change his/her behavior. To be able to correlate this information, a temperature meter and a rain sensor are also used to gather data and to send it to the cloud platform, as showed in Fig. 3. All of the sensors are connected to two NodeMCU ESP8266 (IoT Device).

One microcontroller is continuously gathering data from the door sensor and when a change of state happens, it sends that event to the Edge Hub, via Wi-Fi communication, using a MQTT publish message to a specific topic about the state of the door. The other microcontroller is acquiring information about temperature and rain and sending it periodically to the Edge Hub, via Wi-Fi communication using a MQTT publish message to a specific topic about weather. A RaspberryPi3 model B is responsible to serve as a MQTT broker. An instance of a local Node-RED server (node.js) is also executing on this device. The platform being used is the Google's Cloud Platform. Google develops modules for Node-RED and receives data using MQTT subscribe messages.

This chapter demonstrates the necessary software and hardware that can be used to consort the elder's behavior. A simple set of non-intrusive sensors is used in order to test the communication between sensors and platform. Node-RED is a great tool to help in terms of management because of its graphical interface.

5 Analysis and Discussion

Several tests were performed to achieve the communication between a sensor and Google's Cloud Platform. Considering that a publish/subscribe mechanism is used, scalability is secured. The local network can grow in number of sensors and IoT devices without a significant increase in cloud processing costs, given that the Edge Hub present at the edge of each LAN reduces the data sent to the cloud platform

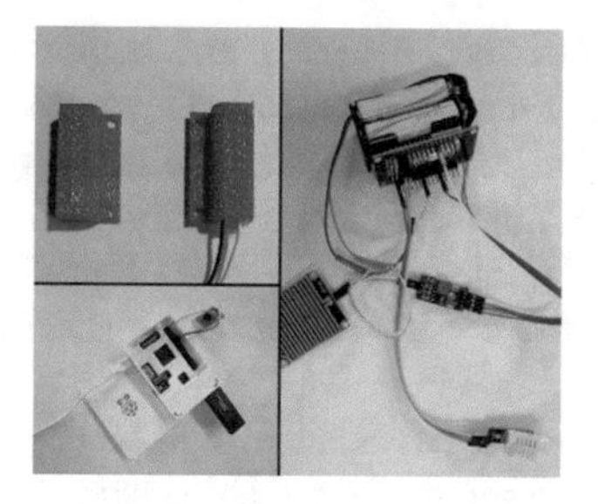

Fig. 3. Magnetic reed switch (top-left), SBC (bottom-left) and weather sensors

regardless of its processing power and storage. This way, through the Edge Hub, each LAN possesses a small part of the responsibility of the AI process, discarding the most unprecedented values. Taking into consideration that the information gathered is received from only non-intrusive sensors, elder's everyday life is not affected by the proposed system. The sensors are "passive" in its method of interaction, and so, elders do not have to interact directly with them. This is important, since this type of technology is not of the common knowledge for the elder. The concept "low-cost" is assured given that communication to the cloud is established with very low-price hardware components and low energy consumption. One set of hardware that includes a magnetic reed switch, an ESP8266 and a RaspberryPi3, can be obtained by values under $50.

A significant amount of data is needed for the algorithm to work at the required precision, so while this data is being generated by the sensors, we are also generating data manually by asking elders to write down their own habits of exiting and entering their homes. Whenever the amount of data needed is fulfilled by sensors themselves, the algorithm will be capable of predicting elder actions in their quotidian life. e.g. "It is usual that the person leaves every day during the week at 3 pm. Given that information, he should leave today as well. If he does not, maybe something is wrong".

In summary, a publish/subscribe mechanism is used to connect the sensors to the platform. Each sensor publishes to a different topic. The broker of this mechanism is the edge hub (RaspberryPi3) which is responsible by handling the communication. In our implementation and testing this device is also responsible for processing information using the Edge A.I. module. When the data reaches the cloud, further AI processing takes place with the aim of understanding and creating patterns of the elder's behavior. Additionally, the elders are asked to provide written information about their habits, given that this type of algorithm requires a large amount of data in order to reach reliable results.

6 Conclusions and Future Work

This paper presents a system being developed with the purpose of helping elders that live alone; this work pretends to provide a high level of security, comfort and tranquility to the person in question, as well as to his/her family and caregivers. The means by which communication between a sensor and a cloud platform is established is detailed; a low-cost and real time communication system is implemented, allowing the platform to receive data from its sensors anytime and anywhere. Data is transmitted through an IoT device, an Edge Hub, and finally to a cloud-based platform. The communication is possible by means of the MQTT protocol. The Edge Hub is presented as the broker, while the IoT device and the platform are its clients as publisher and subscriber.

As future work, it is intended that Edge AI will be on the edge of each LAN (Edge Hub), in order to reduce the amount of data that is sent to the Google Cloud Platform. Considering that a RaspberryPi3 has little processing capabilities, a light algorithm of AI will be presented, in order to discard the most obvious flaws in the data collected regarding elder behaviors. At the platform, a supervised AI algorithm will be developed for achieving more precise results. A security layer will be added to the communications with the purpose of protecting the private and sensitive data that travels through the Internet. At this point, the data travels through the Internet without any encryption. One of the most important/necessary modules to implement is encryption of communication. Very sensitive data is being sent to the cloud. If accessed, a malicious person could have important information about the behavior patterns of the elder. Sensors and houses distinctiveness is not yet tested. However, it could be achieved in a few ways, e.g., by providing each sensor with a unique identification, or by associating each house to a specific topic from the publish/subscribe mechanism.

Acknowledgments. This publication is funded by FCT - Fundação para a Ciência e Tecnologia, I.P., under the projects identified by UID/CEC/04524/2016 and UID/CEC/04524/2019.

References

1. Rica, G., et al.: Being There: Concepts, Effects and Measurements of User Presence in Synthetic Environments. Emerging Communication: Studies in New Technologies and Practices in Communication, vol. 5, pp. 59–82. IOS Press (2003)
2. Rashidi, P., Mihailidis, A.: A survey on ambient-assisted living tools for older adults. IEEE J. Biomed. Heal. Inform. **17**(3), 579–590 (2013)
3. Remagnino, P., Hagras, H., Velastin, S., Monekosso, N.: Ambient intelligence: a gentle introduction. Springer, New York (2005)
4. Shadbolt, N.: Ambient intelligence. IEEE Intell. Syst., 2–3 (2003)
5. Kotsiantis, S.B., et al.: Machine learning: a review of classification and combining techniques. Artif. Intell. Rev. **26**(3), 159–190 (2006)
6. Gazis, V., et al.: Short paper: IoT: challenges, projects, architectures (2015)
7. Gurjar, A., Sarnaik, N.: Heart attack detection by heartbeat sensing using Internet of Things: IoT. Int. J. Mod. Trends Eng. Res. **5**(4), 212–216 (2018)
8. Liang, F., et al.: A survey on the edge computing for the Internet of Things. IEEE Access **6**, 6900–6919 (2017)
9. Blackstock, M., Lea, R.: Toward a distributed data flow platform for the web of things (Distributed Node-RED), pp. 34–39 (2015)
10. Manandhar, S.: MQTT based communication in IoT, May 2017

11. Aran, O., Sanchez-Cortes, D., Do, M.T., Gatica-Perez, D.: Anomaly detection in elderly daily behavior in ambient sensing environments. LNCS, vol. 9997, pp. 51–67. Springer, Cham (2016)
12. Demir, E., Köseoğlu, E., Sokullu, R., Şeker, B.: Smart home assistant for ambient assisted living of elderly People with Dementia. Procedia Comput. Sci. **113**, 609–614 (2017)
13. Dawadi, R., Asghar, Z., Pulli, P.: Internet of Things controlled home objects for the elderly, no. Biostec, pp. 244–251 (2017)

System to Detect and Approach Humans from an Aerial View for the Landing Phase in a UAV Delivery Service

David Safadinho[1], João Ramos[1], Roberto Ribeiro[1], Vítor Filipe[2], João Barroso[2], and António Pereira[1,3(✉)]

[1] School of Technology and Management, Computer Science and Communication Research Centre, Polytechnic Institute of Leiria, Campus 2, Morro do Lena – Alto do Vieiro, Apartado 4163, 2411-901 Leiria, Portugal
davidsafadinho.12@gmail.com, eng.rob.ribeiro@gmail.com, jr.joaoramos@outlook.com

[2] INESC TEC and University of Trás-os-Montes e Alto Douro, Quinta de Prados, 5001-801 Vila Real, Portugal
{vfilipe, jbarroso}@utad.pt

[3] INOV INESC INOVAÇÃO, Institute of New Technologies, Leiria Office, Campus 2, Morro do Lena – Alto do Vieiro, Apartado 4163, 2411-901 Leiria, Portugal

Abstract. The possibility to engage in autonomous flight through geolocation-based missions turns Unmanned Aerial Vehicles (UAV) into valuable tools that save time and resources in services like deliveries and surveillance. Amazon is already developing a drop-by delivery service, but there are limitations regarding the client's id, that can be analyzed in three phases: the approach to the potential receiver, the authorization through the client id and the delivery itself. This work shows a solution for the first of these phases. Firstly, the receiver identifies the GPS coordinates where he wants to receive the package. The UAV flights to that place and tries to locate the receiver on the arrival through Computer Vision (CV) techniques, more precisely Deep Neural Networks (DNN), to continue to the next phase, the identification. After the proposal of the system's architecture and the prototype's implementation, a test scenario to analyze the feasibility of the proposed techniques was created. The results were quite good considering a system to look for one person in a limited area defined by the destination coordinates, confirming the detection of one person with an up to 92% accuracy from a 10 m height and 5 m horizontal distance in low resolution images.

Keywords: Computer vision · Deep Neural Networks · Delivery services · Internet of Things · New generation services · Unmanned Aerial Vehicles

P. Novais et al. (Eds.): ISAmI 2019, AISC 1006, pp. 162–170, 2020.
https://doi.org/10.1007/978-3-030-24097-4_20

1 Introduction

In developed countries there is a solid delivery network assured by specialized companies. However, the volume of products to deliver is continuously increasing due to the growth of online shopping, which justifies the inclusion of autonomous systems to speed up the flow of the packages. The evolution of UAV potentiates the opportunity to fulfill this need of delivering supplies in a fast fashion. Amazon took this fact into account and is currently testing Prime Air, an autonomous drop-by delivery system. Since drones may carry sensible and valuable goods, it is important for the system to release the payload near the person authorized for the reception, instead of dropping the package in a predefined location as the previous service does. So, we can consider three phases in the delivery: the detection of a receiver, the identification for authorization purposes and the landing. In this work, we focus on the detection of the person in order to start the approach and identification of the receiver for authorization purposes.

This work is based on a platform to deploy UAV-based services, already proposed by the same authors in a previous paper [1], where users can control vehicles equipped with multiple and different tools. To improve the mentioned platform, we propose a human detection system based on CV that allows to estimate the location of a person, from an aerial perspective as part of a drone-based delivery service. The objective is to estimate in real-time the landing or delivery site for the drone to get closer and identify the potential receiver as the person who ordered the delivery. This service is intended to be an out-of-the-box solution for people who live remotely and need urgent supplies such as meds or food.

Among the main problems, the camera quality depends on motion blur, prop flicker, UAV vibrations, low resolution and other environmental related limitations (e.g.: solar light reflection, fog, humidity) [2, 3]. The hardware of out-of-the-box drones in the market is usually inaccessible, which means that it isn't possible (conventionally) to embed a visual recognition system among the existent functionalities.

The development of this work results in three main contributions. First, we identify and analyze a problem related to UAV based on new generation services. Then, we study the current CV and IoT related technologies that can be useful for this specific situation and with this information we propose a low-cost system to obliterate the problem. A prototype is evaluated, and the respective results are presented.

The structure of this paper is as follows. In the Sect. 2 we present the state of the art focused on object detection and DNN. Then, in the Sect. 3 we list the requirements and analyze the architecture that supports the proposed solution. In the Sect. 4 the implementation of the prototype is described. Next, in Sect. 5, the test scenario is depicted and finally, the last section includes the conclusions and intended future work.

2 State of the Art

This section includes the presentation of CV-based object detection applications and solutions and the analysis of deep learning algorithms for human detection and possible imagery datasets to be used in the training of the respective models to detect humans.

2.1 Visual Analysis Applications and Solutions

There are multiple situations where visual recognition can be applied. For instance, image classification intends to answer what is depicted in the picture, while object detection answers where. Nowadays there are cloud-based solutions that allow to run the onerous part of the image analysis process in powerful remote computers instead of our devices that can be low-performance. In these cases, a stable and fast internet connection is required to receive the media to interpret it and send the outputs back to the client device. This alternative became popular since state-of-the-art algorithms used for image recognition and object detection based in DNN, a machine learning subset, are onerous.

There are many fields that benefit from visual recognition. In retailing, Amazon Go corresponds to an employee-less store with a technology called "Just Walk Out" that merges computer vision, deep learning and sensor fusion to identify and track the customer around the store and know which products they took from the shelves. In agriculture and botany there are systems to check if the food is proper for consumption (e.g.: spoiled fruit detection) or detect diseases in plants [4]. Related to security and surveillance, there are computer vision systems to recognize license plates or perform facial identification of individuals in crowds, through UAV-based platforms [5]. In autonomous driving and driving assistance, cars are now able to detect obstacles in the road and pavement marks and interpret the road signs (i.e.: image classification) for speed maintenance, for instance [6]. Finally, social platforms like Instagram, Snapchat or Facebook use face detection as an augmented reality technology through the application of interactive virtual graphics over the identified person, commonly called filters.

2.2 Deep Learning, Models and Datasets

Through the last years, many algorithms and techniques have been developed to improve two of the most important performance measures: the accuracy and decrease the resources required to apply them. The last techniques require the creation of a good dataset, which begins with the capture of pictures of objects, that are intended to be detected in the new inputs, in different scenarios. Then, it is necessary to label each picture identifying the location of each object. To overpass this step that is very time consuming, there are public datasets like SceneNet, ImageNet, INRIA Person and Microsoft COCO, where the last two include more people related labels [7, 8]. For instance, COCO integrates 250.000 labels of people.

Recently, a new method to face the object detection and recognition called Convolutional Neural Networks (CNN) was created. In CNN, a specific set of operations is applied to the input to prepare and evidence the most important features in the picture. Considering devices with low computational capabilities, the MobileNet CNN was developed [9]. This network shown better results when used with Single Shot Detector (SSD) algorithm [9] and the combination of these two models constitute a lightweight system for object detection, deployable in small devices like the Raspberry Pi [10].

3 Architecture of the System

This section lists the requirements and the architecture for a solution that considers the detection a potential human receiver from a UAV, when arriving at a pre-defined location with the objective of delivering a package, as part of an autonomous service.

3.1 Requirements

CV plays an important role turning this system user friendly since the receiver should need no equipment (e.g.: smartphone, RFID tag, visual badge) to be identified as the receiver and define the landing site. To perform the detection of the receiver through visual recognition, a camera is required. This camera and the processing equipment need to be low-cost and embeddable in a UAV. In terms of functionality, this system only needs to detect humans from an aerial view. The system should operate moderately fast to scan the area fluidly.

3.2 Specification

This system is made of three modules, embedded in a UAV, that cooperate to make the human detection work as specified, through a camera and deep learning algorithms (CNN), as represented in Fig. 1. The first corresponds to the Navigation Module (A), responsible for autonomously flying the drone to a specified geolocation and perform punctual moves or rotations during the delivery. This autonomy is based on mission planning that allows to set the course of a UAV to a destination through GPS coordinates. Then, we need the Image Acquisition Module (C) that is made of a camera to capture images that are fed to the Processing Module (B), which is responsible for running the machine learning algorithm specified for object detection.

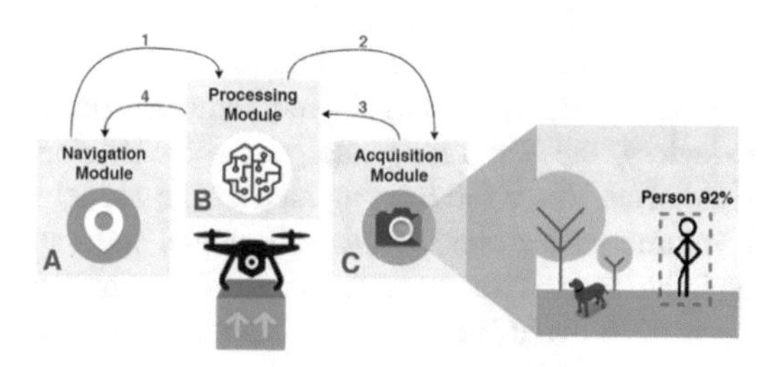

Fig. 1. Architecture of the human detection system.

When the UAV arrives to the GPS coordinates defined in the delivery mission as the receiver's location, the Navigation notifies the Processing (1) to ask the Image Acquisition to turn on the camera (2). The first step corresponds to the acquisition of an image to be posteriorly handled by the Processing (3). Regarding the best-case scenario, the Navigation receives the location of a person to begin the approach and start the next phase of the delivery (4). On the other hand, if this module doesn't detect the presence of a person, the drone will rotate along its vertical axis until a person is detected by the Processing.

4 Prototype

This section describes the development of a prototype and includes the steps we took to get the human detection working on a UAV. Since most out-of-the-box drones available in the market are impossible to program and change through the conventional ways, the UAV to implement this system is assembled modularly, as the one specified in a previous work [11]. The prototype itself is composed by the previously mentioned micro-computer Raspberry Pi, equipped with the Pi Camera v2 with 8MP resolution. This computer is a low-cost solution capable of running a Linux-based operating system, which fulfils the requirements for the installation of OpenCV and TensorFlow.

4.1 Implementation of the System

This system consists in three different modules: navigation, acquisition and processing. The development of the Navigation Module was not taken into account in this prototype, as we only focused on the detection system. The responsibility of this module for this phase is to trigger the Processing Module when the vehicle arrives at the predefined location. At this step, the Processing Module starts the service for the images acquisition, handled by the Acquisition Module.

Relatively to the Acquisition Module, the access to the Raspberry Pi Camera is made with a Python script that uses the OpenCV library to capture images with a 300×300 resolution. Then, we use the same library to manipulate and analyze the images. The first step after the capture is to swap the image color scheme from RGB to a BGR (Blue, Green and Red) matrix due to the specification of the OpenCV library that is expecting images in the same scheme.

About the Processing Module, it is responsible for the person detection algorithm. We chose the best CNN model according to our requirements (i.e.: small-powered ARM architecture device), the SSDLite model, a lighter version of SSD, with the feature-extractor MobileNet. Since the training step of the model requires time and a rich variation and volume of pictures, we used a model pre-trained on the COCO dataset [8] that is available to use directly on the TensorFlow framework. We converted it to the OpenCV format, to avoid the need to run both. After each image processing iteration, the model outputs the prediction for each classification label with its confidence and the location of each object. For this prototype, we are only considering the label "person" with higher confidence than 50%. After more exhaustive tests, this value may be tuned to avoid the detection of non-person objects (i.e.: false positives).

5 Test Scenario and Results

To access the developed solution, we captured 200 photos from an aerial perspective, downscaled them to a square resolution of 300 px, according to the model's specification. In these photos, there is at least one person to be detected. For safety reasons, the photos were captured in a 3-floor building's balconies, to simulate the UAV's AoV.

5.1 Tests Results

In Table 1 we present the results we achieved placing the camera at different horizontal (HD) and vertical distances (VD) from the objects (i.e.: people). The next two columns indicate the used DNN model and the number of samples and next the True Positives (TP), False Positives (FP) and False Negatives (FN) values are presented as well. Finally, the average confidence of the algorithm, the accuracy and the average elapsed time for each image processing were calculated. The TN weren't considered since they represent the absence of people in the images. The algorithm SSDLite was applied to all the VD-HD combinations, while the complete SSD algorithm was only applied to the scenarios with people located further from the camera, in which we expected SSDLite to face more difficulties. The 2nd test scenario, presented in Table 2, minds the characteristics we would expect in a delivery service, meaning only one person to detect from a height of 10 m and horizontal distances of 5 and 10 m, with less noise (i.e.: other objects).

Table 1. Compilation of the tests' results with two individuals.

HD (m)	VD (m)	Model	Samples	TP	FP	FN	Avg. Conf.	Acc.	Avg. time (s)
5	4,5	SSDLite-MN-V2	24	34	0	14	0.7207	0.7083	1.1925
	10	SSDLite-MN-V2	67	68	0	42	0.7230	0.6181	1.1811
10	4,5	SSDLite-MN-V2	45	32	1	53	0.6800	0.3764	1.1877
	4,5	SSD-MN-V2	45	31	3	54	0.6736	0.3647	2.5554
	10	SSDLite-MN-V2	40	3	7	82	0.7386	0.0352	1.1835
	10	SSD-MN-V2	40	14	12	71	0.6359	0.2000	2.472

5.2 Discussion of the Results

By analyzing the average time of processing, we can conclude that the height and distance do not represent noticeable differences in this parameter. The lighter version of the model could finish each step in less than half of the time of the SSD ($\sim$1.18 vs. $\sim$2.5 s).

Table 2. Tests' results from a height of 10 m with just one person.

HD (m)	Model	Samples	TP	FP	FN	Avg. Conf.	Acc.	Avg. time (s)
5/10	SSDLite-MN-V2	24	22	0	2	0.8863	0.9166	1.2608 s

When the camera was placed in a height of 4.5 m and a distance of 5 m to the people (Fig. 2), we could get an accuracy of ~71%. When the camera was at a 10 m height, the accuracy dropped around 9% with a slight decrease on the average time, resulting in an accuracy of ~62%. The value of accuracy was significantly reduced when the people were 10 m apart from the camera. In a 4,5 m height, the accuracy of SSDLite was only ~38%, about ~2% better than the SSD model. Although, from this point we started to get false positives. The false positives found were motorcycles and traffic signals, with confidence around 50%. When the camera was at a 10 m height, the SSDLite got an accuracy of only ~3%, a value far lower than the SSD model, which got an accuracy of 20%. When we compare the average confidence of the true positive values at a 5 m distance, the height didn't make a noticeable difference. Besides, when people were at a distance of 10 m, height had a greater impact. For a height of 4,5 m, both models got the approximate same value (~0,68). From the height of 10 m, SSD got a similar value of ~0,64, while SSDLite got a ~0,74 average confidence. Besides this higher value, it can be explained by the fact that the last model only found 3 true positive individuals whilst the SSD found 14. The difference in the average time between both models only seems to worth the wait when we are in the worst scenario of 10 m height and distance, where SSD improved the accuracy by ~16%. As we will start the search for people in pre-defined GPS coordinates that have a low error margin, we won't expect this situation to happen in a real delivery case. From the 24 samples we gathered on the scenario with a single person with less noise (i.e.: different objects), the algorithm could detect the person in 22 of them, even if they had some occlusions, representing an accuracy of ~92%.

Fig. 2. Person detection result from a 4,5 m height and 5 m distance.

6 Conclusions and Future Work

The problem analyzed in this work consists in the lack of flexibility in the choice of the landing site, detection and authentication of the receiver in a UAV-based delivery service. However, this document focusses in the first step: detection of the potential receiver through an embedded human detection CV system. To deal with the identified problems, we propose an architecture, based on a modularly built quadcopter, that is divided in three modules that help estimating the location of the receiver to begin the approach. We tested the proposal with a simple prototype developed with a Raspberry Pi 3 and a camera. After the implementation we created a test scenario with two main variables: height and horizontal distance. The SSDLite model performed quite well in all the scenarios except for the distance and height of 10 m, where it only got an accuracy of $\sim$4%. We tested the same scenario with the SSD, improving the accuracy by $\sim$16% while doubling the elapsed time between iterations. In a scenario with only one person we reached an accuracy of $\sim$92%. On distant scenarios, the output started to show false positive detections on motorcycles and traffic signals.

As future work, we pretend to implement the complete delivery process. The first task should consider that if the vehicle doesn't find a human, it must at least rotate its body or camera to scan the area around. The next step of the delivery service is the authorization through facial recognition and the final one consists in the interaction of the receiver with the drone, through movement patterns. Regarding the performance of the human detection system, the creation of a bigger dataset including a wider diversity of human morphology and poses in order to create a more accurate model is one of the future tasks.

Acknowledgments. This work is financed by National Funds through the Portuguese funding agency, FCT - Fundação para a Ciência e a Tecnologia within project: UID/EEA/50014/2019.

References

1. Safadinho, D., Ramos, J., Ribeiro, R., Caetano, R., Pereira, A.: UAV multiplayer platform for real-time online gaming. AISC, vol. 571, pp. 577–585 (2017)
2. Beleznai, C., Steininger, D., Croonen, G.: Multi-modal human detection from aerial views by fast shape-aware clustering and classification. In: 2018 10th IAPR Workshop on Pattern Recognition in Remote Sensing, no. i, pp. 1–6 (2018)
3. Xu, S., Savvaris, A., He, S., Shin, H.S., Tsourdos, A.: Real-time implementation of YOLO +JPDA for small scale UAV multiple object tracking. In: 2018 International Conference on Unmanned Aircraft Systems, ICUAS 2018, pp. 1336–1341 (2018)
4. Mohanty, S.P., Hughes, D.P., Salathé, M.: Using deep learning for image-based plant disease detection. Front. Plant Sci. **7**, 1–10 (2016)
5. Motlagh, N.H., Bagaa, M., Taleb, T.: UAV-based IoT platform: a crowd surveillance use case. IEEE Commun. Mag. **55**(2), 128–134 (2017)
6. Ranft, B., Stiller, C.: The role of machine vision for intelligent vehicles. IEEE Trans. Intell. Veh. **1**(1), 8–19 (2016)
7. Taiana, M., Nascimento, J.C., Bernardino, A.: An improved labelling for the INRIA person data set for pedestrian detection. Lect. Notes Comput. Sci. (including Subser. Lect. Notes Artif. Intell. Lect. Notes Bioinformatics), vol. 7887, pp. 286–295. Springer, Heidelberg (2013)
8. Lawrence Zitnick, C., Lin, T.-Y., Maire, M., Belongie, S., Hays, J., Perona, P., Ramanan, D., Dollár, P.: Microsoft COCO: Common objects in context. In: Proceedings of IEEE Computer Society Conference on Computer Vision and Pattern Recognition, pp. 3686–3693 (2014)

9. Howard, A.G., et al.: MobileNets: efficient convolutional neural networks for mobile vision applications (2017)
10. Zhang, Y., Bi, S., Dong, M., Liu, Y.: The implementation of CNN-based object detector on ARM embedded platforms. In: Proceedings of the IEEE 16th International Conference on Dependable, Autonomic and Secure Computing, IEEE 16th International Conference on Pervasive Intelligence and Computing, IEEE 4th International Conference on Big Data Intelligence and Computing and Cyber Science and Technology Congress, vol. 50, pp. 379–382 (2018)
11. Safadinho, D., Ramos, J., Ribeiro, R., Pereira, A.: UAV proposal for real-time online gaming to reduce stress (2018)

Smart Coach—A Recommendation System for Young Football Athletes

Paulo Matos[1,3](✉), João Rocha[1], Ramiro Gonçalves[2], Ana Almeida[1,3], Filipe Santos[1,3], David Abreu[1], and Constantino Martins[1,3]

[1] Computer Science Department, Engineering Institute - Polytechnic of Porto, Porto, Portugal
{psm,jsr,amn,jpe,1140272,acm}@isep.ipp.pt
[2] Trás-os-Montes e Alto Douro University, Vila Real, Portugal
ramiro@utad.pt
[3] GECAD Knowledge and Decision Support Research Group, Porto, Portugal

Abstract. Over the last decades Information and Communication Technologies (ICTs) are increasingly being used in sports, especially in football, aiming to improve the athletes training and results. However, training systems for young athletes do not have, for the most part, learning abilities in order to adapt, evolve and find new training recommendations, designed specifically for each young athlete.

In this paper introduce the *Smart Coach* user adaptation model, and whose main goal is to present our hybrid recommendation system to help young athletes evolve. This facilitate the interaction between members of a club technical staff and their young athletes, reinforcing the young person counselling, and their potential as an athlete.

Keywords: Recommender systems · User modelling · Personalized coaching

1 Introduction

Information and Communication Technologies (ICTs) are increasingly being used in the world of sport, especially in football, aiming to enhance athletes training methods, improve the team results or support sports decisions and refereeing.

However, training systems for young (semi)amateur athletes do not, for the most part, consider their performances and accomplishments in training and competition, regarding the young athletes' characteristics, technique, tactics, physical and mental status, in the training selection and recommendation process.

On the other hand, these systems do not have learning capabilities in order to adapt, evolve and find new training recommendations for each young person. These limitations, make the results of these systems, not adapted, and not focused on the players specificities.

P. Novais et al. (Eds.): ISAmI 2019, AISC 1006, pp. 171–178, 2020.
https://doi.org/10.1007/978-3-030-24097-4_21

It is in this context that the *Smart Coach* recommendation system, intends to innovate and make impact. Using artificial intelligence technology and techniques, to support coaches and technical staff, allowing them to analyse better their young athletes skills and to enhance their development and training [5,8].

In this paper we introduce the user adaptation model of *Smart Coach*, targeting the evolution of young athletes. In summary, *Smart Coach* will allow to represent technical, tactical, physical and/or psychological characteristics of young athletes, and adapt a Dynamic Training Model, defining a training schedule to improve a young sportsman performance, targeting is evolution as a player.

In Sect. 2 we make a brief description of User Modelling and Recommendation Systems, assessing some applications developed specifically for football coaching. In Sect. 3 we describe in detail the *Smart Coach* proposed architecture and in Sect. 4 we take some conclusions and present the future work.

2 State of Art

In this section we resume the current user modelling and recommendation systems state of art. There is also a review of five football coaching applications.

2.1 User Modelling

User modelling is normally implemented with two sets of techniques, the knowledge-based and the behavioural [7]. Knowledge-based adaptation typically result from information gathered using forms, queries and other user studies, with the purpose to produce a set of heuristics. Behavioural adaptation is related with user monitoring during his daily tasks and activities [11].

In the last decades, several different systems were developed to store various kinds of user information. Some of those applications were analysed and reviewed in works done by Morik, Kobsa, Wahlster and McTear in 2001. In those first systems, user modelling was an application part, which caused difficulties to separate users profile related processes from the other application components. This was a normal problem in software design before software encapsulation and modularization techniques became popular. Despite the technology evolution related with users profiles modelling (it has become more complex and intelligent, by making use of newly technological evolutions), the basic concepts and ideas and problems that turned possible the appearance of this research area are almost the same: the identification of user needs, desires, personalities and, most important, objectives.

Different user modelling techniques and methodologies were used to represent knowledge, some of them are data representation oriented and others data inference oriented [12]. The User Modelling techniques (linear models, decision trees, neural networks, text mining, Bayesian networks and data mining) are all forms of predictive statistical models, since they are applied in areas with thousands or millions of items (from products, clients, actions, etc.) and can also benefit from recent machine learning evolution [5,8,17]. Finally, not all of them might actually be applied in some domains, due to their specific characteristics [12].

Linear models are probably the most common techniques, and it can probably even be said that almost all systems uses linear models, one way or another. These models are easy to build and understand; they are efficient and assume probabilistic data as believable effects, which has been a successfully employed theory so far [17].

2.2 Recommendation Systems

A Recommender System can be characterized like a collection of different techniques used by different systems to filter and organize its items in order to select either the best ones or the most suitable ones for presentation, according to the user [10]. Although the most common scenario is when the system has to choose the best items from a certain group which otherwise (without the filtering) would be randomly selected, there are other more important cases where certain items or types of items just can't be shown to the user at a given moment, for example, due to player field position. A complete recommender system should therefore be prepared to handle both types of situations. The mode of operation normally used by recommender systems is to use a knowledge base (the user model) as the basis for a series of calculations to infer which are going to be, amongst all the items available, the ones that will better please the user, according to a wide variety of theories or approaches [11]. In this work is considered that the best way to please users is to suggest trainings that can improve their abilities and their game-play insufficiencies.

Recommend something to someone carries an implicit responsibility to whom does that, because it is fundamental to assure accuracy and quality in the recommendation results in order to gain users confidence. These systems are basically based in three types of paradigms (content, collaborative and knowledge-based) and all their possible combinations [4,6,8,13]. Content-based filtering tries to capture information from within the content of unstructured or disorganized item data elements, such as textual or descriptive attributes, generally including powerful text mining algorithms from the information retrieval area. Collaborative filtering (also called social-filtering) is one of the currently most used techniques and was greatly influenced by the Web 2.0 ("social web") phenomena. It relies on other user's information for recommending items to the current user [2]. Knowledge-based filtering is almost inevitable to use, because it means using any form of domain knowledge in a recommender system [12].

In some systems the referred techniques are combined to take advantage of each approach characteristics and also mitigate limitations. The systems are characterized as hybrid approaches [5].

2.3 Football Coaching Software

In this section we describe five applications used to help football coaches manage their teams and briefly explain their capabilities. None of these applications include any artificial intelligence mechanism namely recommendation systems.

Tactical Soccer—Is a desktop/mobile commercial application, with capabilities to manage football teams and single players. Can be used to create exercises, manage training sessions, define match formations, line-ups and strategies. This software was developed with the main objective of improving quality and productivity in individual and collective training [16].

MyCoach Football—Is a free desktop/mobile application, intended to amateur/professional football coaches. The application can manage team and match information, create and organize trainings. Another advantage of this software is the ability to share exercises and training session information with the community of coaches using the application. There is also a mobile application, that can be used to collect match information [9].

Dossier do Treinador—Is a commercial web application focused in football and futsal. The software includes a complete exercise editor, automatic statistics, player profiles, calendar and attendance management, and generation of several reports (e.g. game reports, calendars, player profiles) [3].

SportEasy—Is a free/commercial web and mobile application, used to manage amateur teams. Can be used in any sport and can be accessed by all members of the club. The software entitles himself as "the best way to manage your sports team" but lacks training and exercises management capabilities [15].

Soccer Coach - Team Sports Manager—Is a free/commercial mobile application (exclusive for IOS devices) meant for football coaches, offering the capability to manage teams, players, define training sessions and manage exercises from a list of available exercises (conditioned by the used version)[14].

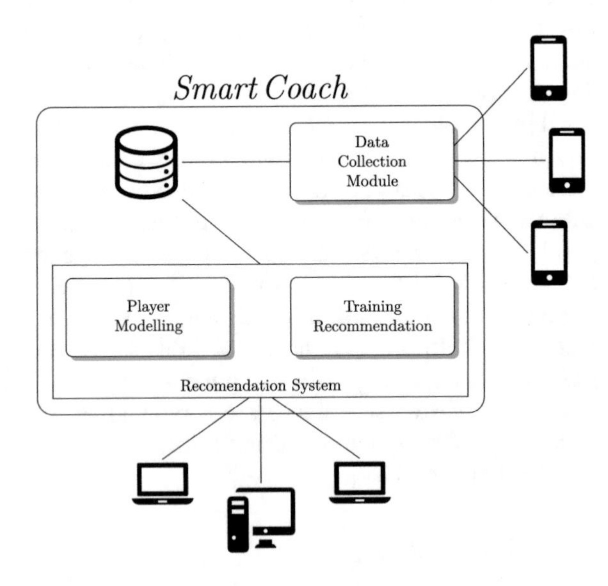

Fig. 1. Diagram of proposed approach

3 Proposed Approach

This paper proposes the creation of a training recommendation system for young football players, based on athletes performance during matches. This performance information, is collected by friends/family using a web responsive application and stored in a database for future use by the recommendation system (see Fig. 1).

User profiles will be created based on players position, physical characteristics (*e.g.* height, weight, speed, jumping height, etc.) and match performance. The Table 1 shows the most important attributes collected for each young player during matches. Some of this attributes are more or less important, depending the position the athlete plays during matches.

The player modelling module, presented in Sect. 3.1, creates a player profile based in all the data previously collected. In conjunction with the training recommendation module, filters and selects the recommended training for that specific athlete. This recommendation process is presented at the end of Sect. 3.1, and illustrated with Fig. 2.

Table 1. Performance attributes to collect during matches for players and the importance by position

Attribute	Goalkeeper	Centre Back	Full-back/Wing-back	Defensive Midfielder	Centre midfield	Attacking midfield	Winger	Centre forward
(In)Complete saves	•	–	–	–	–	–	–	–
Passing accuracy	•	•	•	•	•	•	•	○
Clearances	○	•	•	•	•	○	○	○
(In)Complete interception	○	•	•	•	○	○	○	○
Ball recovery	–	○	○	•	•	•	○	○
(In)Successful tackles	–	•	•	•	•	•	○	○
Fouls committed	•	•	•	•	•	•	•	•
Fouls suffered	○	○	○	○	•	•	•	•
(In)Successful dribbles	–	○	○	○	○	•	•	•
Duels won/lost	–	•	•	•	•	•	•	•
(In)Successful crosses	–	○	•	○	○	•	•	○
Shots/Shots on target	–	○	○	○	•	•	•	•
Offsides	–	○	•	○	•	•	•	•
Assists	–	•	•	•	•	•	•	•
Goals	–	•	•	•	•	•	•	•

• Major attribute ○ Minor attribute – Not applicable

3.1 Smart Coach User Profile Modelling

The *Smart Coach* young athletes User Modelling solution to be implemented, implied initially the athletes' stereotype definition. These stereotypes, were defined during interviews with several football coaches, and the K-Means [1] clustering algorithm was applied to data obtained during matches of two Portuguese football academies and collected according to attributes defined in Table 1. Each cluster was classified with a set of attributes with diverse weights and mapped according to their relevance in training, and is team performance impact. The clustering outcome are the young football players profiles, *e.g.*, goalkeeper, defender, midfielder, striker. Each cluster has a user type, which is classified with several kinds of attributes/tasks that typically is performed during a football match.

The characteristics that define each young athlete, his particularities and origins, have influence in is performance during a football match, and can define is game-play approach (*i.e.* a more or less aggressive posture, a technical or a physical player, etc.). These goals are directly extracted from the domain model and define the user domain-dependent data. Each of these stereotypes corresponds to a set of objectives, tasks or functions. The athletes goals are reached, when they successfully complete a set of actions, necessary to the conclusion of certain training.

In the user model, each training (regardless of granularity) has associated a performance percentage (scale of 0 to 10) that allows the confirmation of training successful completion. This value is used in the adaptation model (interaction) to allow recommendation adjustment for trainings suggested to staff and young athletes (Fig. 2). As a result, based on his personal performance/training history, the user experience is improved with suggestions for specific workouts (Fig. 2).

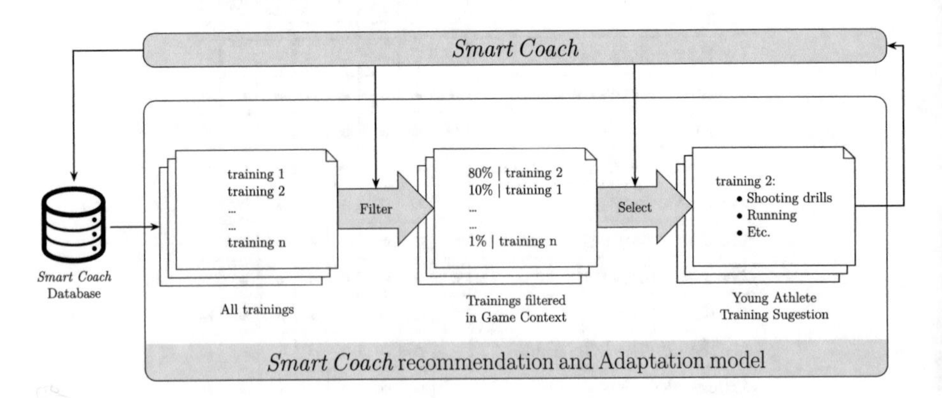

Fig. 2. Smart Coach recommendation and adaptation model

3.2 Training Determination Process

The training determination process intends to infer, based on the young athlete state captured at a certain point of time, the set of trainings the athlete is performing in the same time window (Fig. 2).

This information allows the determination of relevant workouts appropriate to actions being performed. Determining what action is being performed, is an uncertain and imprecise process. The observation of athletes' actions may not have a objective interpretation, and as such, may result in erroneous conclusions.

In the *Smart Coach* system context, the relationship between training exercises will have parameter elements associated with the young athlete model stereotypes. These elements relate to the young athletes' performance parameters function, such as the training and the necessary steps to successful complete them. Thus, for each stereotype a hierarchical set is defined, that associates all the trainings that are within its competence and the relations between them are represented. For representation, a directed acyclic graph is used.

In view of the initial catalogue of all possible training sessions, the adaptation model (Fig. 2) will first use information about the user's stereotypes and attributes, considered relevant by the staff to perform/apply our filtering algorithm. This will allow for example, that only trainings related to young athletes' actions and goals are considered. Hence, an ordering is made according to the probability of each activity being performed at a given time, using data such as, the expected course of training and the young athletes' personal performance history. In a next step, the young athlete performance information (considering the static data *Smart Coach* collected) will allow to select the volume, training type and respective information of the exercises the young athlete should perform (Fig. 2).

4 Conclusion and Future Work

The first version of the data collection prototype, is being evaluated in two football clubs' academies, which we will designate by club A and club B.

The study will run over the course of two months with two teams: one from club A and one from club B. The athletes age is between 15 and 18 years.

None of the athletes or staff had any experience in using mobile applications to support sport. However, everyone is familiar with mobile applications and personal computers (PCs), namely, surfing the Internet, using social networking frequently and playing video-games.

All data is being stored in a Microsoft Excel file, and then imported into a statistical data processing application, the IBM Statistical Package for the Social Sciences (SPSS).

Acknowledgements. This work was supported by the Project 2017-1-PT01-KA202-035903, From Birth to Adult Age—a WBL successful Practice, co-funded by the ERASMUs+ programme of the European Union and by National Funds through the FCT—Fundação para a Ciência e a Tecnologia (Portuguese Foundation for Science and Technology) within the Project UID/EEA/00760/2019.

References

1. Anderberg, M.R.: Cluster analysis for applications. Technical report, Office of the Assistant for Study Support Kirtland AFB N MEX (1973)
2. Berka, T., Plößnig, M.M.: Designing recommender systems for tourism. Proceedings of ENTER 2004, pp. 26–28 (2004)
3. Dossier do treinador de football (2018). http://www.dossierdotreinador.com/
4. Felfernig, A., Gordea, S., Jannach, D., Teppan, E., Zanker, M.: A short survey of recommendation technologies in travel and tourism. OEGAI J. **25**(7), 17–22 (2007)
5. Isinkaye, F., Folajimi, Y., Ojokoh, B.: Recommendation systems: principles, methods and evaluation. Egypt. Inform. J. **16**(3), 261–273 (2015)
6. Kabassi, K.: Personalizing recommendations for tourists. Telematics Inform. **27**(1), 51–66 (2010)
7. Kobsa, A.: Generic user modeling systems. User Model. User-Adap. Inter. **11**(1–2), 49–63 (2001)
8. Lu, J., Wu, D., Mao, M., Wang, W., Zhang, G.: Recommender system application developments: a survey. Decis. Support Syst. **74**, 12–32 (2015)
9. My coach football - the digital assistant for educators (2018). https://www.mycoachfootball.com/en/
10. Porter, J.: Watch and learn: How recommendation systems are redefining the web, the Internet, **5** (2006). https://articles.uie.com/recommendation_systems/. Accessed 13 Dec 2006
11. Santos, F., Almeida, A., Martins, C., Gonçalves, R., Martins, J.: Using poi functionality and accessibility levels for delivering personalized tourism recommendations. Comput. Environ. Urban Syst. (2017). https://doi.org/10.1016/j.compenvurbsys.2017.08.007, http://www.sciencedirect.com/science/article/pii/S0198971517302880
12. Santos, F., Almeida, A., Martins, C., Oliveira, P., Gonçalves, R.: Tourism recommendation system based in user functionality and points-of-interest accessibility levels. In: Mejia, J., Muñoz, M., Rocha, Á., San Feliu, T., Peña, A. (eds.) Trends and Applications in Software Engineering, pp. 275–284. Springer International Publishing, Cham (2017)
13. Schafer, J.B., Konstan, J., Riedl, J.: Recommender systems in e-commerce. In: Proceedings of the 1st ACM Conference on Electronic Commerce, pp. 158–166. ACM (1999)
14. Soccer coach | the definitive coaching app. (2018). http://www.teamsportsmanager.com
15. Online sports team management software - sporteasy (2018). https://www.sporteasy.net/en/home/
16. Tactical soccer - complete suite for soccer coaches (2018). https://tacticalsoccer.co.uk/
17. Zukerman, I., Albrecht, D.W.: Predictive statistical models for user modeling. User Model. User-Adap. Inter. **11**(1–2), 5–18 (2001)

Functional Prototype for Intrusion Detection System Oriented to Intelligent IoT Models

Jose Aveleira-Mata[1(✉)] and Hector Alaiz-Moreton[2]

[1] Research Institute of Applied Sciences in Cybersecurity (RIASC) MIC, Universidad de León, 24071 León, Spain
jose.aveleira@unileon.es

[2] Department of Electrical Engineering and Systems and Automatic, Universidad de León, León, Spain
hector.moreton@unileon.es

Abstract. The importance of IoT (internet of things) systems, that allow things to be connected to the Internet and increase their functionalities, is becoming increasingly more relevant. The number of connected devices is growing exponentially. The special features of these devices, and the protocols used in IoT systems, make them more vulnerable to intrusion attacks. New needs arise in terms of network security. To improve the security of an IoT system without affecting the performance of the systems, an IDS (Intrusion Detection Systems) is proposed to detect anomalies in the IoT environment. In order to do so, machine learning techniques as well as the dataset used and the classification method must be taken into account. Our research focuses on the development of an IDS prototype that takes the network frames of an IoT environment using the MQTT protocol, a dataset with a compilation of attacks in a system that uses the protocol, and tests a classification model in a real time environment.

Keywords: IoT · Intrusion Detection Systems · MQTT protocol · IDS

1 Introduction

The IoT (internet of things) is the central technology for transforming houses and cities into smart houses and cities. It is based on the concept of connecting all things to the internet. The 2017 report from the International Data Corporation [1] predicts that 50 billion new devices will be connected by 2020. Gartner published a similar study [2] in which, according to Cisco, 500 billion devices are expected to be connected to the Internet by 2030. Each device includes sensors that collect data, interact with the environment and communicate over a network. The Internet of Things (IoT) is emerging as a business concept among the world's principal IT companies such as Microsoft (Azure IoT), CISCO (Cisco IoT for business), Google (google home ecosystem), Amazon (Alexa with IoT consumer products), etc. [3].

In this new context, it is possible for objects, services and applications to make decisions and react according to a given situation in their environment, for example to control the temperature and consumption of home electricity in order to optimize its use and improve comfort.

P. Novais et al. (Eds.): ISAmI 2019, AISC 1006, pp. 179–186, 2020.
https://doi.org/10.1007/978-3-030-24097-4_22

However, major security concerns arise with the adoption of IoT. As a result of the increasing complexity of IoT networks, these networks use a wide variety of protocols [4, 5] and different devices that usually have low computing capacities to reduce cost and energy requirements.

Attacks in IoT are possible as the systems in the IoT network are an easy target to hack [6, 7], And the IoT devices have no virus or malware protection software. This is a natural consequence of the low-memory and low-power nature of these devices. One of the most significant attacks in recent years, the Mirai botnet, carried out Distributed Denial of Service (DDoS) attacks by exploiting these vulnerabilities in a series of IoT network devices infected with malware, known as bots or zombies, attacking as many as 400,000 simultaneously connected devices [8].

One way to manage network security is the use of (IDS) Intrusion Detection Systems. These systems use anomaly detection techniques on network traffic to detect intrusions within the network [9]. The use of IDS is a viable approach for the security of IoT systems as it does not require more computing capacity in the devices, and its detection rules can be adapted according to the requirements of the protocols.

In order to improve IDS systems, Machine Learning models are proposed which, after being trained with normal and under-attack frames [10], can detect anomalies. In this way it would be possible to adapt the new IoT protocols and networks in IDS.

Our work consists of the development of a prototype of IDS that selects the characteristics of the most relevant traffic frames according to the protocol and the attack to be detected, passing this information to the models in real time to detect different attacks, in this specific case on the MQTT protocol which is widely used in IoT [11], in such a way that its efficiency can be tested in an environment with real traffic.

2 Related Work

There are multiple datasets to analyze the anomalies. These datasets collect network data with normal and under-attack frames, one of the most popular is the Kdd99 [12] and its improved version, the NLS-KDD [13], it contains different attacks such as Neptune-DoS, Pod-DoS, SmurfDoS, and buffer-overflow, Other interesting datasets for the detection of anomaly datasets containing botnet attacks, such as the UNSW-NB15 [14]. An interesting and current dataset is the AWID [15] dataset which consists of the collection of Wi-Fi network traffic frames with several attacks. In our case, due to the lack of public datasets with network data in IoT environments, we will use our own dataset that collects network data from an IoT system that we will see in the next section.

As regards datasets, it is interesting to make a selection of the most relevant features when detecting attacks, since this way the detection of anomalies is improved by taking only the most important features into account, to reduce the characteristics of the KDD dataset. Researchers use then feature selection method based on the lightning attachment procedure optimization algorithm (LAPO) [16] with good results in experiments into efficiency and accuracy in searching for the optimal feature subset. The same dataset also proposes the use of a hybrid system using IGR techniques together with the k-means [17] algorithm with which it is possible to identify which sort of attack is registered in the dataset. (D-FES) are used for the detection of Impersonation attacks [18], and Weighted-Feature Selection for the Neural Networks Classifier [19] in the WIDs deep-feature extraction and selection techniques.

The IDS use machine learning classification techniques to classify the frames that may be under attack such as support vector machines (SVM), K-nearest neighbor (k-NN), Artificial neural networks. [20] used primarily for anomaly detection systems (ADSs) that are based on network anomalies and can detect a greater variety of attacks than a rule-based IDS [21].

As regards IDS oriented to the IoT, recent research uses KDD attacks that can also affect IoT systems such as User to Root (U2R) and Remote to Local (R2L) attacks. By proposing models using component analysis and the linear discriminate analysis of dimension reduction [22], AD-IoT proposes a detector that uses the UNSW-NB15 dataset that contains the most modern network attacks of 2015 with a conceptual framework [23]. This dataset together with simulated data from IoT sensors is used to extract the proposed features and evaluate the ensemble technique to analyze botnet attacks in particular against DNS, HTTP and MQTT protocols used in IoT networks [24].

From the IDS for the IoT [25, 26] reviews it can be deduced that improvements in these approaches would be to use datasets directly taken from IoT environments and be able to validate their operation in real scenarios.

3 Methods and Materials

This section details how the functional prototype of IDS has been created for machine learning models for the IoT as the elements that make up the system.

3.1 The Dataset

The dataset collects information from an environment using several sensors and actuators that use the MQTT protocol together with a server that manages the protocol, and other computers that consult the status of the devices. This dataset consists of three files containing the frames collected from three attacks, Mitm: contains 110,668 frames with 3,855 under man-in-the-middle attack and 106,813 normal traffic frames; Dos: contains the capture of 94,625 frames of which 45,513 are under attack traffic and 49,112 are normal traffic; Intrusion: contains 80,893 total frames with 1,898 under attack and 78,995 normal traffic frames. All the frames are tagged as frames under attack or normal.

(The dataset is available in https://joseaveleira.es/dataset. © reg#LE-229-18)

3.2 Classification Method

The classification method used XGBoost with a selection of the most relevant characteristics from the dataset that has had good results in a previous research [27], as can be seen in Table 1.

Table 1. Results of evaluation metrics for XGBoost.

Model	M. logarithmic loss	M. classification error rate
XGBoost-Train	0.075348	0.024753
XGBoost-Test	0.079451	0.025651

A model has been created with it which is used within the server that analyses the network frames. Each one of the frames passes through the model to be tagged in real time.

3.3 Development of the Prototype Intrusion Detection System

Router Configuration. We have used a low-cost, TP-link router with the OpenWrt Linux operating system [28], which allows us to configure the port mirroring and thus get to send copies of all network packets seen in the Wi-Fi interface (wlan0).

In order to change the configuration of the router to carry out the port mirroring, it is necessary to have the system updated and install the iptables-mod-tee [29] package and change the configuration to redirect traffic to our server by configuring the iptables rules to mirror upstream and downstream traffic.

Server Development. The server has been developed in Python to facilitate the upload and use machine learning models. It has been developed with the flask library [30] as this allows us to create a simple API REST that can grow in functionalities.

We capture traffic from the port mirroring to the server running using the 'tshark' Wireshark tool [31] with which we can capture traffic through the command line with the option of selecting the fields of the frames we want to capture, thus we can dissect the traffic to capture the most relevant characteristics of the frames, and adapt it to different protocols and IoT networks.

The dissected and formatted frames separated by commas (CSV) which we then pass directly to the model which classifies and tags them as normal frames or under attack. Finally, we use Websockets to send this information to a client application that shows whether an intrusion has taken place in real time.

User Interface Development. We have developed a UI in Vue.js to show the traffic [32] We have chosen this technology so that the prototype because it provides modular development in JavaScript. We can grow in functionality through new components. So far we developed the functionality of the user login and monitoring of the frames displaying the relevant features of the frames and the label of it are under attack or not. Showing this information consumes the WebSocket provided by the server.

IDS system. As can be seen in Fig. 1. the IDS collect all the frames of the WLAN network and processes them through a machine learning model.

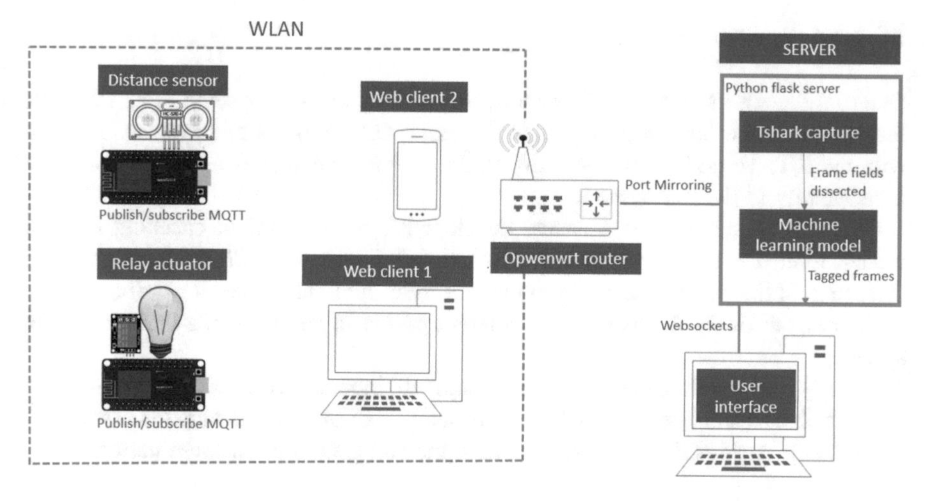

Fig. 1. Intrusion detection system schema

4 Experiments

We capture the traffic of the system seen in Fig. 1. using the prototype IDS. We then capture the traffic in real time so that the model classifies the frames. There is a certain delay so that it can be visualized in the user interface in a correct format. We insert a time delay of 0.14 s between each frame publication.

```
socketio.emit('frame', stream + tag)
eventlet.sleep(0.14)
```

The IDS dissects the frames to select the fields necessary to classify the MQTT protocol, sends each frame to the model, which is trained with an XGBoost classifier to detect DoS attacks to the MQTT server.

From a computer in the WLAN we implement a DOS Attack and from the user interface it can be seen that the frames that are under attack are tagged as "intrusion", as seen in Fig. 2.

FRAMES EXIT

Frames

ip.src	tcp.srcport	eth.src	MQTT.msgtype	type
216.214.174	443	18:a6:f7:eb:77:26		normal
216.214.174	48168	4c:74:03:07:92:4f		normal
216.214.174	48168	4c:74:03:07:92:4f		normal
192.168.171	1883	30:5a:3a:62:72:80		INTRUSION
192.168.171	1883	30:5a:3a:62:72:80	beem.loadr-malaria-VirtualBox-5638-0	INTRUSION

Fig. 2. Intrusion frames in the user interface

5 Conclusions

Due to the great growth of IoT systems, the security of these systems is an important factor, one of the biggest problems is the variety of protocols and systems that arise with the IoT. To address this security IDS can detect anomalies at the network level transparently to the IoT system.

An IDS prototype that we have developed is very useful in checking how the models trained in a real system with new network traffic, and with the functionality of dissection of the frames according to the most relevant fields makes it possible to adapt to the new needs for the analysis of different protocols and networks that require IoT systems.

The system is still under development and will include new improvements in future work, such as testing other models and the analysis of other types of networks used by IoT systems such as ZigBee and Bluetooth including sniffers for these networks.

References

1. Barakat, S.M.: Internet of Things: Ecosystem and Applications. J. Curr. Res. Sci. **4**(1), 32–34 (2017)
2. Mohn, E.: Internet of Things. Salem Press Encyclopedia of Science (2018)
3. Lueth, K.L.: The top 20 Internet of Things companies right now, iot-analytics.com, (2015). https://iot-analytics.com/20-internet-of-things-companies/. Accessed 11 Jan 2019
4. Chernyshev, M., Baig, Z., Bello, O., Zeadally, S.: Internet of things (IoT): Research, simulators, and testbeds. IEEE Internet Things J. **5**(3), 1637–1647 (2018)
5. Granjal, J., Monteiro, E., Sa Silva, J.: Security for the Internet of Things: a survey of existing protocols and open research issues. IEEE Commun. Surv. Tutorials **17**(3), 1294–1312 (2015)
6. Neshenko, N., Bou-harb, E., Crichigno, J., Kaddoum, G., Ghani, N.: Demystifying IoT security: an exhaustive survey on IoT vulnerabilities and a first empirical look on internet-scale IoT exploitations, pp. 1–30, April 2019
7. Ahmad, M., Younis, T., Habib, M.A., Ashraf, R., Ahmed, S.H.: A review of current security issues in Internet of Things. In: Jan, M.A., Khan, F., Alam, M. (eds.) Recent Trends and Advances in Wireless and IoT-enabled Networks, pp. 11–23. Springer, Cham (2019)
8. Kolias, C., Kambourakis, G., Stavrou, A., Voas, J.: DDoS in the IoT: mirai and other botnets. Comput. (Long. Beach. Calif) **50**(7), 80–84 (2017)
9. Garcia-Teodoro, P., Diaz-Verdejo, J., Maciá-Fernández, G., Vázquez, E.: Anomaly-based network intrusion detection: Techniques, systems and challenges. Comput. Secur. **28**(1), 18–28 (2009)
10. Al-Mandhari, I.S., Guan, L., Edirisinghe, E.A.: Investigating the effective use of machine learning algorithms in network intruder detection systems. In: Advances in Information and Communication Networks, pp. 145–161 (2019)
11. Sethi, P., Sarangi, S.R.: Internet of Things : Architectures, Protocols, and Applications, vol. 2017 (2017)

12. Tavallaee, M., Bagheri, E., Lu, W., Ghorbani, A.A.: A detailed analysis of the KDD CUP 99 data set. In: IEEE Symposium on Computational Intelligence in Security and Defense Application CISDA 2009, no. Cisda, pp. 1–6 (2009)
13. Aggarwal, P., Sharma, S.K.: Analysis of KDD dataset attributes - class wise for intrusion detection. Procedia Comput. Sci. **57**, 842–851 (2015)
14. Moustafa, N., Slay, J.: UNSW-NB15: a comprehensive data set for network intrusion detection systems (UNSW-NB15 network data set). In: Proceedings of 2015 Military Communications and Information Systems Conference MilCIS 2015, pp. 1–6 (2015)
15. Kolias, C., Kambourakis, G., Stavrou, A., Gritzalis, S.: Intrusion detection in 802.11 networks: empirical evaluation of threats and a public dataset. IEEE Commun. Surv. Tutorials **18**(1), 184–208 (2016)
16. Sun, S., Ye, Z., Yan, L., Su, J., Wang, R.: Wrapper feature selection based on lightning attachment procedure optimization and support vector machine for intrusion detection. In: Proceedings 2018 IEEE 4th International Symposium Wireless Systems Within International Conferences Intelligent Data Acquisition Advanced Computing Systems, pp. 41–46 (2018)
17. Araújo, N., De Oliveira, R., Ferreira, E., Shinoda, A.A., Bhargava, B.: Identifying important characteristics in the KDD99 intrusion detection dataset by feature selection using a hybrid approach. In: 2010 17th International Conference on Telecommunications, ICT 2010, pp. 552–558 (2010)
18. Aminanto, M.E., Choi, R., Tanuwidjaja, H.C., Yoo, P.D., Kim, K.: Deep abstraction and weighted feature selection for Wi-Fi impersonation detection. IEEE Trans. Inf. Forensics Secur. **13**(3), 621–636 (2017)
19. Aminanto, M.E., Tanuwidjaja, H.C., Yoo, P.D., Kim, K.: Wi-Fi intrusion detection using weighted-feature selection for neural networks classifier. In: 2017 International Workshop on Big Data and Information Security, pp. 99–104 (2017)
20. Tsai, C.F., Hsu, Y.F., Lin, C.Y., Lin, W.Y.: Intrusion detection by machine learning: a review. Expert Syst. Appl. **36**(10), 11994–12000 (2009)
21. Moustafa, N., Hu, J., Slay, J.: A holistic review of network anomaly detection systems: a comprehensive survey. J. Netw. Comput. Appl. **128**, 33–55 (2019)
22. Pajouh, H.H., Javidan, R., Khayami, R., Ali, D., Choo, K.-K.R.: A two-layer dimension reduction and two-tier classification model for anomaly-based intrusion detection in IoT backbone networks. IEEE Trans. Emerg. Top. Comput. **6750**(c), 1 (2016)
23. Alrashdi, I., Alqazzaz, A., Aloufi, E., Alharthi, R., Zohdy, M., Ming, H.: AD-IoT: anomaly detection of IoT cyberattacks in smart city using machine learning. In: 2019 IEEE 9th Annual Computing and Communication Workshop and Conference, pp. 0305–0310 (2019)
24. Moustafa, N., Turnbull, B., Choo, K.K.R.: An ensemble intrusion detection technique based on proposed statistical flow features for protecting network traffic of internet of things. IEEE Internet Things J. **PP**(c), 1 (2018)
25. Benkhelifa, E., Welsh, T., Hamouda, W.: A critical review of practices and challenges in intrusion detection systems for IoT: toward universal and resilient systems. IEEE Commun. Surv. Tutorials **20**(4), 3496–3509 (2018)
26. da Costa, K.A.P., Papa, J.P., Lisboa, C.O., Munoz, R., de Albuquerque, V.H.C.: Internet of Things: a survey on machine learning-based intrusion detection approaches. Comput. Netw. **151**, 147–157 (2019)

27. Alaiz-Moreton, H., Aveleira-Mata, J., Ondicol-Garcia, J., Muñoz-Castañeda, A.L., García, I., Benavides, C.: Multiclass classification procedure for detecting attacks on MQTT-IoT protocol. Complexity **2019**, 1–11 (2019)
28. openwrt.org. https://openwrt.org/. Accessed 15 Jan 2019
29. iptables-mod-tee. https://openwrt.org/packages/pkgdata/iptables-mod-tee. Accessed 05 Feb 2019
30. flask. http://flask.pocoo.org/. Accessed 06 Feb 2019
31. wireshark.org. https://www.wireshark.org/docs/man-pages/tshark.html. Accessed 05 Feb 2019
32. vuejs.org. https://vuejs.org/. Accessed 06 Feb 2019

Workshop on Ambient Intelligence for e-Healthcare (AIfeH)

Workshop on Ambient Intelligence for e-Healthcare (AIfeH)

Scope

The world healthcare market is living an increasing demand towards e-Healthcare services. E-Healthcare is a stimulating and inspiring area of research as providing solutions to many different problems and needs. We can find e-Healthcare systems that help to improve mental health by detecting people moods and emotions; e-Healthcare systems that are used for providing medical services; e-Healthcare systems that are applied to treat mobility problems, and so on. E-Healthcare enlists electronic processes and communications to provide quality healthcare services to patients, covering all locations.

The diversity of problems and needs turns the development of e-Healthcare systems into a challenging activity requiring disciplines as different as Artificial Intelligence, Human-Computer Interaction and Engineering to work together in order to provide solutions that satisfy the growing demand of the society. The stakeholders, including companies and industry, must be able to create systems that are as intelligent as to detect falls or emotions, as usable as to be exploited by people with mobility problems, and as reliable and autonomous as to monitor health and mobility.

Topics

This workshop focuses on current achievements and future trends of information and communication technologies, methodologies and applications to solve problems of real life related to e-Healthcare focusing on the following (but not limited) topics:

Smart environments in support of e-Healthcare.
Innovative Information Technologies for the development of inclusive e-Healthcare systems.
Requirements engineering in product modelling of e-Healthcare systems.
Novel architectures in the development of e-Healthcare systems.
New frameworks to support the development of e-Healthcare systems.
Robots and agents integration in the development of e-Healthcare systems.
Novel sensor infrastructures in the development of e-Healthcare systems.
Case studies of e-Healthcare systems.

Committee

Organizing Committee

Antonio Fernández-Caballero	Universidad de Castilla-La Mancha, Spain
Elena Navarro	Universidad de Castilla-La Mancha, Spain
Pascual González	Universidad de Castilla-La Mancha, Spain

Program Committee

Jose Carlos Castillo	Universidad Carlos III de Madrid, Spain
Javier Jaen	Universitat Politécnica de Valencia, Spain
S. Shyam Sundar	Sungkyunkwan University, Korea
Albert Salah	Bogazici University, Turkey
Jose Manuel Pastor	Universidad de Castilla-La Mancha, Spain
Jesus Favela	CICESE, Mexico
Sylvie RattÈ	Ecole de technologie supÈrieure, Canada
Enzo Pasquale Scilingo	University of Pisa, Italy
José Ramón Álvarez-Sánchez	Universidad Nacional de Educación a Distancia, Spain
José Ramón Álvarez-Sánchez	Universidad Carlos III de Madrid, Spain
Rafael Martínez-Tomás	Universidad Nacional de Educación a Distancia, Spain
Ilias Maglogiannis	University of Piraeus, Greece
Julie Doyle	Dundalk Institute of Technology, Ireland
Bogdan Kwolek	AGH University of Science and Technology, Poland

Continuous Authentication in Mobile Devices Using Behavioral Biometrics

Rodrigo Rocha[1], Davide Carneiro[1,2(✉)], Ricardo Costa[1], and César Analide[2]

[1] CIICESI/ESTG - Polytechnic Institute of Porto, Felgueiras, Portugal
{8140411,dcarneiro,rcosta}@estg.ipp.pt
[2] Algoritmi Centre/Department of Informatics, Universidade do Minho, Braga, Portugal
analide@di.uminho.pt

Abstract. In recent years, the development and use of mobile devices such as smartphones and tablets grew significantly. They are used for virtually every activity of our lives, from communication or online shopping to e-banking or gaming, just to name a few. As a consequence, these devices contribute significantly to make our lives more digital, with all the perks and risks that this encompasses. One of the most serious risk is that of an authorized individual gaining physical access to our mobile device and, potentially, to all the applications and personal data it contains. Most of mobile devices are protected using some kind of password, that can be easily spotted by unauthorized users or event guessed. In the last years, new authentication mechanisms have been proposed, such as those using traditional biometrics or behavioral biometrics. In this paper we propose a new continuous authentication mechanism for mobile devices based on behavioral biometrics that monitors user interaction behavior for classifying the identity of the user.

Keywords: Continuous authentication · Behavioral biometrics · User classification

1 Introduction

In the last years, mobile devices have become increasingly important in our daily living. Their development in terms of processing power, storage, communication and mobility led to an unprecedented growth in existing mobile applications, that now cover virtually the whole spectrum of our daily living, from leisure activities to work. As a consequence, we now move around with a small portable device in our pockets that contains very sensitive information that ranges from our social networks to online shopping or e-banking accounts. [2] shows that 40% of the users stores information that they consider *secret* on their mobile device, and that nearly one third of mobile users had accessed at least once a smartphone that was not theirs, such as a lost or stolen one. When our smartphone is accessed by a non-authorized individual, the user gains potential access to all this sensitive information [6].

P. Novais et al. (Eds.): ISAmI 2019, AISC 1006, pp. 191–198, 2020.
https://doi.org/10.1007/978-3-030-24097-4_23

Indeed, when a smartphone is stolen or lost, the value lost is well beyond the value of the device itself. In an experiment conducted by the security company Symantec, 50 smartphones with no authentication mechanisms were left in five North-American cities. Data was collected regarding if and how they were used. Results show that 96% of the devices were accessed by someone. Of these, in 86% of the cases the unauthorized user accessed personal information and, in 60% of the cases, this information included social networks and e-mail accounts [9].

The most commonly approach used to secure mobile devices is information-based authentication, such as the use of text-based passwords (e.g. a numeric code) or a pattern that is drawn on the screen. More recently, with the availability of better/cheaper sensors and computational powers, biometric authentication mechanisms have been used (e.g. face recognition, fingerprint). However, and as will be addressed in Sect. 2, these methods still have disadvantages. Namely, the former can be easily spotted by an authorized user and reproduced for gaining physical access to device in order to access its information. The latter (or even both) are irrelevant if the unauthorized user is able to gain access to the device while it is unlocked, as happens if it is stolen while being used by the owner. In this paper we propose a continuous authentication mechanism that is based on the interaction behavior of the user of the device, and on an interaction profile built for each authorized user. The approach continuously monitors user interaction and locks the device whenever a significant difference exists between the profile and the current interaction.

2 Existing Authentication Mechanisms

As previously addressed, there is the need to effectively secure mobile devices, given the amount of personal and sensitive information about their users that they hold nowadays. The most common solutions for this are the use of text-based passwords, PIN numbers or patterns (the user draws a specific figure on the screen, using a predetermined set of points). The main advantage of these methods is that they are very easy to use and require very few computational resources. However, it is also very easy for an unauthorized user to observe the owner of the device while unlocking the device, namely in public transports, restaurants or even in places where there is video surveillance [10]. Moreover, most users tend to use passwords that are easy to memorize. These are, often, also the passwords that are easier to guess. Indeed, it has been shown that in 9.23% of the times that an unauthorized user gains access to a device by "password-guessing", it does so in less than three attempts [7]. This happens in part because most users define their password based on important dates (such as birth date) or on repetitions of a single number such as 0000 or 1111. The use of patterns for unlocking the screen is sometimes also relatively easy to guess as these often leave markings on the screen of the device [10]. More advanced techniques even thermal imaging cameras to accomplish this same goal [1].

More recently, authentication methods based on biometrics have emerged that promise increased security. These can be divided in two main categories: physiological and behavioral.

Physiological biometrics authentication mechanisms rely on the use of specific sensors, such as fingerprint scanners or cameras for facial recognition. This has as main disadvantage the increase of the cost of the device. It has also been shown that some of these approaches can be fooled, namely by using photographs of people [3].

Behavioral biometrics, on the other hand, relies on behaviors of the user such as specific actions, application usage, habits, among others. In this kind of authentication, the behavior of the user is monitored in search of actions or habits that are generally not attributed to the user.

These methods can also be characterized as explicit or implicit (or continuous). Explicit methods require the input of some form of authentication on a specific moment (e.g. fingerprint, password). The device remains unlocked afterwards. Implicit or continuous authentication methods continuously monitor the device and the user in search for signs that the current user is not the authorized one. Some example of these methods can be found in the literature. In [4], the authors use specific gestures on the screen (e.g. flick, drag, pinch) and a specifically designed glove to identify the user. In [5] the authors propose the so-called Typing Authentication and Protection, that relies on the user's virtual key dynamics for identification.

In this paper we propose a continuous authentication mechanism based on the user's interaction with the device. As opposed to existing methods, it does not require specific hardware nor gesture recognition: it is based on a group of features that are extracted from the simple interaction events with the screen of the smartphone, and that are not application-dependant.

3 Methodology

3.1 Data Collection

To validate this approach, an experiment was setup in which users interacted regularly with their mobile devices, throughout the day. During this time, the mobile application collected data about each touch performed with a single finger on the screen. In order to reduce the variability of the data, touches are aggregated using a first-in first-out sliding window of size 30. The average values of each variable in the sliding window are considered to build the interaction features.

Specifically, twelve interaction features are considered. Nine of them are the minimum, maximum and average values of touch duration, area and intensity. The remaining three are obtained by considering touch duration over time and fitting a quadratic curve to the data. This quadratic curve represents the general touch pattern of the user, being a composite feature that combines the duration and intensity features. From this curve we obtain the remaining three variables, which are the three coefficients of the quadratic fit, which we deem x^2, x and n, respectively. This data is aggregated and processed locally in the device, and is then sent to a central server where it is stored.

Data was collected from 30 users, 15 women and 15 men, with ages ranging between 10 and 67 ($\bar{x} = 35.95$, $\sigma = 14.96$). This resulted in the collection of 1665 instances of data (each representing the result of aggregating over a sliding window of 30 touches). After collection, data was normalized.

A comparative analysis of the data for each user was conducted, to determine if there are significant differences. As shown in Figs. 1 and 2, there are indeed differences between users. Figure 1 shows how average touch duration is distributed for each user, showing significant differences between the median value as well as the standard deviation.

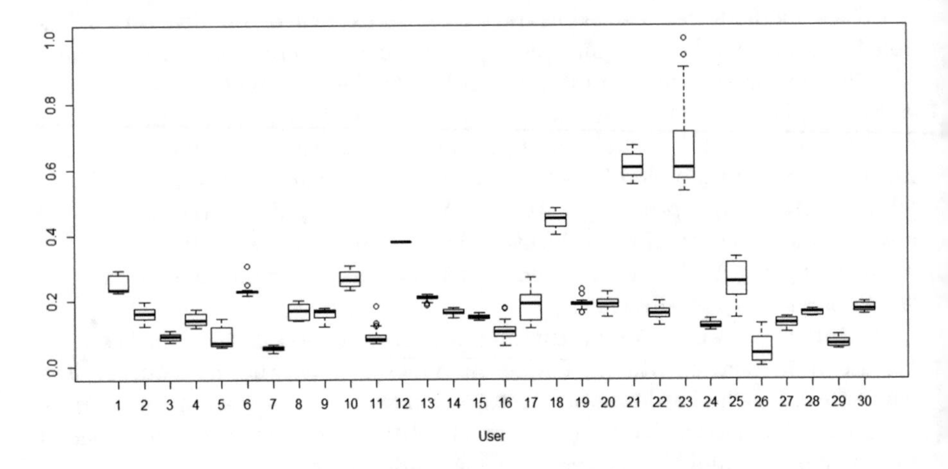

Fig. 1. Distribution of average touch duration for all the users.

Similarly, Fig. 2 shows the distribution of touch intensity and, once again, the differences between each user are clearly visible. Similar differences are observed in the remaining variables. This preliminary analysis supports the assumption followed in this paper that, when considering all the variables, their distribution should be unique for each user, such as her/his fingerprint. Figure 3(a) shows the average value of each variable for 3 different users, graphically depicting these differences in each one's interaction patterns with the mobile device.

3.2 User Modeling and Classification

Having determined that there are, indeed, differences between the interaction patterns of users, work proceeded towards the creation of an *interaction model* that could effectively be used for identity classification. To this end, the following approach was implemented. For each user and for each variable we calculated the four quartiles and the interquartile range (IQR). Based on this value, we defined the *normal* upper and lower limits for each feature/user. These limits

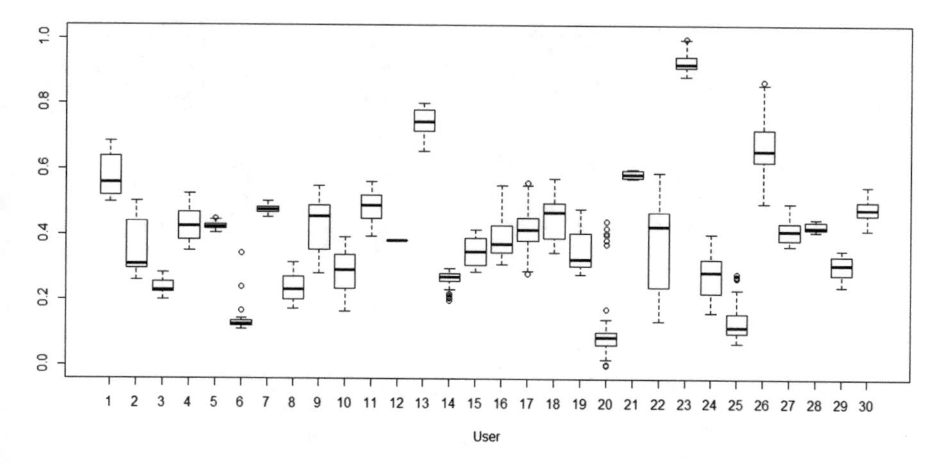

Fig. 2. Distribution of average touch intensity for all the users.

represent the boundaries between which each user generally behaves while interacting with the device. They were defined in Eqs. 1 and 2, as proposed by John Tukey (who first proposed boxplots) and is commonly used in these diagrams as a demarcation line for outliers [8].

The interaction profile of each user is updated at regular intervals, when new interaction data is available in the central server. Profiles are stored in a central database, in which there is one instance for each user. Each instance of a profile contains 26 variables: the identifier of the user, the 24 limits for the twelve features (upper and lower) and the timestamp in which the profile was last updated. This database is then used by the classification service to determine the identity of a given user.

$$lower_{u,i} = Q1_{u,i} - 1.5 * IQR_{u,i} \tag{1}$$

$$upper_{u,i} = Q3_{u,i} + 1.5 * IQR_{u,i} \tag{2}$$

Figure 3(b) represents graphically this interaction profile for User 1. The upper and lower limits are represented in a solid line (red and green, respectively). Two instances of data are also represented, one from the same user (dotted blue line) and another from a different user (dashed black line). The picture also shows that the dotted line is completely within the boundaries of the user's interaction profile while the dashed line is only partially within those limits.

The user classification service thus works as follows. The user accepts requests containing an instance of processed interaction data and the identifier of the owner of the device in which the data was collected. The service then retrieves that user's interaction profile from the database and proceeds to compare the values of the features in the interaction instance received with that interaction profile. Specifically, it compares the value of each feature against the corresponding upper and lower limits in the interaction profile, and calculates the percentage

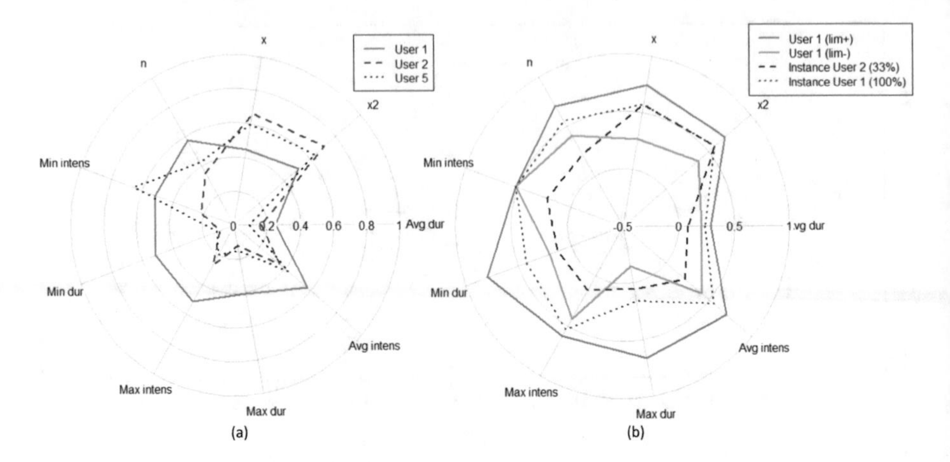

Fig. 3. (a) Graphical representation of the average values of each interaction feature, for three users. (b) Graphical representation of the interaction profile of User 1, and comparison with two instances for classification, one from the same user and another from a different one.

of features that fall within the limits. Then, if this percentage is above a specific threshold, it is assumed that the current user of the device is the authorized user. Otherwise, it is assumed that an unauthorized user gained access to the device.

Mobile devices thus constantly send interaction data to this central server. If the data are classified as belonging to the authorized user, it is added to the central database and later to the user profile. On the other hand, if the data is classified as not belonging to the owner of the device, the mobile device may choose to lock the screen and immediately request an explicit authentication mechanism, such as a pattern or a fingerprint. If the user successfully unlocks the device, the misclassification is considered a false positive and the instance of data is saved in the central server nonetheless, associated to the owner of the device. If, on the other hand, the user fails to unlock the device, the breach is validated and it is considered a true positive. In this scenario, the interaction data is not saved in the central server as it represents data from a different user.

3.3 Experimental Results

In order to assess the suitability of the proposed approach, the following approach was implemented. The data was split into two groups: the train split containing 75% of the data (1248 observations) and the test split containing the remaining 25% (417 observations). Interaction models for the thirty users were built following the methodology described in Sect. 3.2 and using the data in the train split.

Then, the test split was further processed to simulate authentication breaches. Indeed, in our dataset all data was gathered without the occurrence of unauthorized accesses. Thus, to simulate these attacks, 67% of the instances were changed to simulate that they belong to a user different than the actual user. The remaining 33% were left unchanged. Each instance of data was then submitted to the classifier service, which classified it as *breach* (positive) or *no breach* (negative). Given that the dataset contains information about the actual user that produced the interaction data, it is possible to calculate the confusion matrix, as depicted in Table 1.

Table 1. Confusion matrix for the prediction of an identity breach.

		Actual	
		Yes	No
Pred.	Yes	266	6
	No	20	125

In the table, 266 instances represent true positives (actual breaches that were detected), 125 represent true negatives (cases in which the data belonged to the owner of the device and the service classified it accordingly), 6 represent false positives (the service determined that it wasn't the owner of the device interacting when it actually was) and 20 represent false negatives (the service predicted that it was not a breach when it actually was). The results show an accuracy of 93.76% (precision = 97.79%, recall = 93.01%).

4 Conclusions and Future Work

In this paper we presented an approach for continuous authentication in mobile devices using behavioral biometrics. We extract twelve interaction features from the user's interaction with the screen of the smartphone. These features describe the user interaction in different modalities including time, intensity and area, and combinations of these. Based on these features we build individual interaction profiles that can be used to classify the identity of a user given an instance of interaction with a satisfying accuracy.

Based on this, we developed an online service that continuously collects data about the users' interaction with the devices, and that updates interaction profiles at regular intervals. This service is then used by mobile applications to continuously verify the identity of the user, in a non-intrusive way.

In future work we will continue to collect data from more users, to determine how the accuracy of the service changes with an increased number of user profiles. We are also implementing and endpoint for client applications to download the model, so that classification can be carried out locally, without requiring a constant data connection. We will also study and quantify the relevance of each variable towards user identification, so that the service can be improved to use a weighted sum in order to improve accuracy. Given the presented results, we believe that the proposed interaction features can effectively be used in a real-life authentication application, that is lightweight, continuous and non-intrusive.

Acknowledgments. This work has been supported by FCT – Fundação para a Ciência e Tecnologia within the Project Scope: UID/CEC/00319/2019.

References

1. Abdelrahman, Y., Khamis, M., Schneegass, S., Alt, F.: Stay cool! understanding thermal attacks on mobile-based user authentication. In: Proceedings of the 2017 CHI Conference on Human Factors in Computing Systems, pp. 3751–3763. ACM (2017)
2. Boyles, J.L., Smith, A., Madden, M.: Privacy and data management on mobile devices. Pew Internet American Life Project 4 (2012)
3. Edmunds, T., Caplier, A.: Motion-based countermeasure against photo and video spoofing attacks in face recognition. J. Vis. Commun. Image Represent. **50**, 314–332 (2018)
4. Feng, T., Liu, Z., Kwon, K.A., Shi, W., Carbunar, B., Jiang, Y., Nguyen, N.: Continuous mobile authentication using touchscreen gestures. In: 2012 IEEE Conference on Technologies for Homeland Security (HST), pp. 451–456. Citeseer (2012)
5. Feng, T., Zhao, X., Carbunar, B., Shi, W.: Continuous mobile authentication using virtual key typing biometrics. In: 2013 12th IEEE international Conference on Trust, Security and Privacy in Computing and Communications (TrustCom), pp. 1547–1552. IEEE (2013)
6. Jain, A.K., Shanbhag, D.: Addressing security and privacy risks in mobile applications. IT Prof. **14**(5), 28–33 (2012)
7. Lu, L., Liu, Y.: Safeguard: user reauthentication on smartphones via behavioral biometrics. IEEE Trans. Comput. Soc. Syst. **2**(3), 53–64 (2015)
8. Tukey, J.W.: Exploratory Data Analysis, vol. 2. Reading, Mass (1977)
9. Wright, S.: The symantec smartphone honey stick project. Symantec Corporation, March 2012
10. Zhou, L., Kang, Y., Zhang, D., Lai, J.: Harmonized authentication based on thumbstroke dynamics on touch screen mobile phones. Decis. Support Syst. **92**, 14–24 (2016)

Improving Motivation in Wrist Rehabilitation Therapies

Miguel A. Teruel, Víctor López-Jaquero, Miguel A. Sánchez-Cifo, Elena Navarro(✉), and Pascual González

LoUISE Research Group, Computing Systems, University of Castilla-La Mancha, Albacete, Spain
{MiguelAngel.Teruel, VictorManuel.Lopez, Elena.Navarro, Pascual.Gonzalez}@uclm.es

Abstract. Rehabilitation encompasses a wide variety of activities aimed at reducing the impact of injuries and disabilities by applying different exercises. Frequently, such exercises are carried out at home as a repetition of the same movements or tasks to achieve both motor learning and the necessary cortical changes. Although this increases the patients' available time for rehabilitation, it may also have some unpleasant side effects. That occurs because carrying out repetitive exercises in a more isolated environment may result in a boring activity that leads patients to give up their rehabilitation. Therefore, patients' *motivation* should be considered an essential feature while designing rehabilitation exercises. In this paper, we present how we have faced this need by exploiting novel technology to guide patients in their rehabilitation process. It includes a game crafted to make recovery funny and useful, at the same time. The game and the use we made of the specific hardware follow the recommendations and good practices provided by medical experts.

Keywords: Wrist rehabilitation · Myo · Motivation

1 Introduction

Rehabilitation is defined by the World Health Organization [21] as an "active process by which those affected by injury or disease achieve a full recovery or, if a full recovery is not possible, realize their optimal physical, mental and social potential and are integrated into their most appropriate environment". Rehabilitation encompasses a wide variety of activities intended to reduce the impact of injuries and disabilities by applying different processes and strategies.

For a proper rehabilitation process, designing or selecting the most appropriate exercises for each patient is needed, as well as to provide a suitable feedback. It is frequent that such exercises embrace the repetition of the same movements or tasks to achieve both motor learning and the necessary cortical changes [13]. Moreover, patients are usually requested to carry out part of such exercises at home, without the therapist's supervision or feedback [22]. Although this increases their time available for rehabilitation, it may also have some unpleasant side effects, because carrying out repetitive exercises in a more isolated environment, such as their home, without

P. Novais et al. (Eds.): ISAmI 2019, AISC 1006, pp. 199–206, 2020.
https://doi.org/10.1007/978-3-030-24097-4_24

interacting with other peers or therapists may result in a boring activity that leads patients to give up their rehabilitation [11]. Thus, patients' *motivation* should be considered an essential feature while designing rehabilitation exercises in order to increase the chances of achieving a successful rehabilitation. The use of technology, as Langan [16] pointed out, may be used as one of the cornerstones for improving motor performance, as well as engaging patients by means of entertainment and the use of objective feedback. In this paper we present how we have exploited a cheap device, namely Myo armband [29], for wrist rehabilitation. It uses an electromyography (EMG) sensor to collect muscular activity data from the forearm, facilitating gesture recognition such as grasp gestures, finger gestures [12] or a combination of them. This device is used in the context of an application to monitor patients' movements and evaluate whether they are achieving their expected goals or not.

This paper is structured as follows. After this introduction, some of the main facts about wrist rehabilitation are introduced in order to present the foundations of this work. Then, in Sect. 2 the related work is presented. In Sect. 3 the software developed is described, highlighting the motivational aspects it features. Lastly, in Sect. 4, the main conclusions drawn with this work are presented.

2 Related Work: Electromyograph for Monitoring Wrist Motion

The human wrist is a complex structure composed of different elements that enables a person to adapt the hand orientation depending on the task performed. Moreover, it allows locking the hand in certain moments when required, e.g. picking up objects or making gestures [25]. Focusing on specific motion of the wrist, there are two main axes and a mixture of single rotations in which movement occurs. Rotation around the forearm axis is not possible but is accomplished with specific movements of some elements forming the wrist [15]. The measurement of active wrist range of motion supplies information on the muscle's ability to produce and transfer force, and the condition of structures around joints of the body [14]. Our lifestyle makes a wrist with a wide range of movement necessary for our daily tasks, as even basic ones in which hands are implied require slight motion, e.g. pointing at one direction keeping the hand forming a straight plane with the forearm. Additionally, the wrist provides precision when pointing because it can be moved a few degrees in any axis.

Research in rehabilitation has focused on finding new training procedures which emphasize on how to perform movements, including intensity, repetition and activity, such as [5, 31]. In [10, 18], research is based on task specific techniques involving one-on-one interaction with a therapist who encourages the patient [6]. Techniques are not standardized, and measurements may vary, due to the usage of distinct practices among clinics [14]. Generally, the exercises recommended by therapists usually encompass active and passive wrist flexion, extension, pronation and supination. Depending on the level of damage, the degree of movements vary and the help received by customized molds change with the patient's necessities [26].

When it comes to recovering the wrist from injuries or surgery, and going deep into the specific recommended exercises, doctors focus treatment on the following basics [30] for a faster recuperation: bend the hand towards floor and then raise up towards the ceiling, ensuring that the movement is only performed by the wrist; locate the hand upwards, then turn the palm down and up repeatedly, keeping the elbow still; pose the hand forming a straight position with the forearm, afterwards tilt the hand one way and then the other; make a tight fist, ensuring that knuckles are bent to a right angle and help with the other hand.

The need of human motion tracking in the rehabilitation domain is not new [32]. During decades different devices for tracking human movements have been presented. These devices provide real-time data that dynamically captures the pose changes of a human body. But now, the emergence of cheap devices, such as Leap Motion [17], a marker-free visual based tracking systems, and Myo armband [29], an electromyography (EMG) sensor that collects muscular activity data in the forearm, provide new possibilities to include them in at-home wrist and hand rehabilitation. In this context, although there are proposals that combine the use of these two devices [23], they are limited by the space tracking constraints that Leap Motion exhibits, as its effective range extends from approximately 25–600 mm above the device (1 inch–2 feet). Thus, to overcome this limitation, other types of devices can be integrated that do not suffer from this shortcoming. In particular, we can use those based on the analysis of EMG signals that capture electrical activity of contracting muscles.

Although, the use of electromyography (EMG) for controlling prostheses and rehabilitation is not novel [8], now with the emergence of cheap EMG devices new opportunities to introduce their massive use are opened. As we have pointed out, one of these devices is Myo armband, developed by Thalmic Lab. It can capture EMG data via eight stainless steel superficial electrodes. Moreover, it has nine-axis inertial measurement sensors, provides haptic feedback and has Bluetooth communication. Within the constraints in terms of its limited bandwidth (<200 Hz), there are studies [20] which, after comparing the results with other EMG, state that its limited bandwidth is not necessarily a drawback in detecting different movements.

Myo is broadly used in hand gesture recognition, including grasp gestures [2], finger gestures [12] or a combination of wrist and finger gestures [1]. Moreover, there are studies that analyze the feasibility of using Myo for understanding one's arm movements during the physiotherapy stage [24]. Likewise, there are proposals that include hand gesture recognition and serious games, whose objective is to improve the engagement and motivation of the patients. After presenting some relevant aspects that should be taken into account while designing motor rehabilitation games, Batista et al. [3] presents a game based on Simon game board, where the user should replicate some gestures in a defined order. Other proposal [4] presents a new version of breakout Atari game that makes use of Myo armband for controlling the movement. Moreover, in [27] a training game is presented that aims to setup prosthesis wearers for success by mapping typical controls used for prostheses to game input. Its goal is not only to train to regain lost functionality, range-or-motion, but also to learn how to manage a new prosthesis. In general, all of these proposals attempt to include new algorithms to detect gestures from Myo signals. It is not so common to find proposals that use directly the gestures recognized by Myo, introducing some motivating aspects in its design, while providing a mechanism to the therapist to design a specific wrist rehabilitation therapy.

3 Myo-Rehab: Motivating Wrist Rehabilitation Exercises

Myo-Rehab is the tool developed to perform wrist rehabilitation therapies in a motivating manner. With this aim, we applied gamification [9], motivation [19] and awareness [28] principles to create an enjoyable and engaging experience.

The game procedure is quite straightforward; as soon as the game begins and the Myo armband is synchronized with the computer, the player will be asked to perform a series of exercises. Such exercises consist in performing certain exercises and repeating them for a number of times that the therapist determined. Such exercises are *extension*, *flexion*, *pronation*, *supination*, *grip strengthening* and *finger spring*. It is worth noting that, in order for these movements to be properly recognized, the Myo armband can be calibrated in advance by reproducing each of the aforementioned exercises.

When the player is required to perform an exercise, the game will show an image representing how to perform such exercise (Fig. 1(a)). Moreover, during each repetition of an exercise, the player will be asked to perform the exercise and, when the system detects that such exercise has been performed, it will ask the player to put the wrist in the original resting position. With this aim, the messages *Do it!* and *Rest...* will be presented to the player both visually and aurally. Furthermore, the player will receive haptic feedback from the armband after completing each repetition.

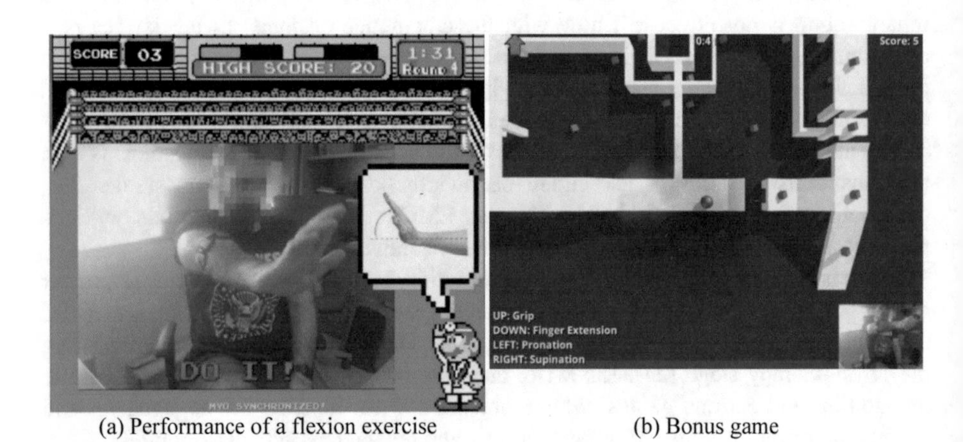

(a) Performance of a flexion exercise (b) Bonus game

Fig. 1. Look and feel of Myo-Rehab

Each therapy will have a time limit, established by the therapist, for the player to complete it. Hence, if the time is over, the game will finish. However, if the player finishes the game and there is any remaining time, he/she will play the bonus level. Such level will differ from the previous game considerably. In this case, the idea is to focus on the player enjoyment rather than in the rehabilitation process, yet the latter is considered in a different manner. Thus, in the bonus game, the player will control a ball that he/she will have to move around a labyrinth to touch and destroy several explosive red cubes (Fig. 1(b)). To move the ball, the player will have to perform a movement with his/her wrist corresponding to the previous ones. However, in this case, instead of doing an exercise and resting repeatedly, the exercise position is maintained to move the ball in one direction.

As it can be seen in the left-bottom corner of Fig. 1(b), the four directions the ball can be moved in will correspond with the mentioned exercises. Moreover, the correspondence among exercises and ball movements can be customized by the therapist for each player. Furthermore, while playing this level, feedback will be provided to the player by using red arrows (top-left corner of Fig. 1(b)), that will make the player aware of the exercise that is being performed at any time.

3.1 Motivation for Myo-Rehab

As aforementioned, it is not enough to implement a digital platform enabling a patient to do the exercises he/she does in a clinical rehabilitation environment at home. On the contrary, the social facet of rehabilitation, where the patient interacts with both other patients and therapists must be reproduced or replaced with artefacts that prevent the user from giving up the rehabilitation because of the lack of motivation produced by the isolation when the patient does rehabilitation at home. Neglecting the consideration of motivational features in computer-aided rehabilitation environments can result in patients giving up rehabilitation. In Myo-Rehab this issue has been carefully addressed from two points of view: gamification and influence.

Gamification relies on the use of video game elements in non-gaming systems [9]. To catch the attention of the user with the gamification part of the exercise, it has been designed following the aesthetics of 8-bit video games from the 80 s. The purpose of choosing this aesthetics is two-folded. First, this type of low-resolution graphics design sends a clear message of playing a game, and thus provides a friendly informal environment in which rehabilitation therapy can be embedded. Second, the gaming elements introduced in the environment are simple enough to avoid distracting the user from the main purpose of the application: rehabilitation.

Choosing a boxing video game was decided to use some recurring metaphors in such game genre that match more formal terms used in rehabilitation. During the performance of the therapy, its status will be presented in gamified style. Hence, each performed repetition of the exercise will increase the player's score, as it happens in any video game when a goal is achieved. Moreover, his/her high score will also be shown, thus influencing the player to try to surpass it. Besides, taking advantage of the game theme, namely a boxing combat, the performance of each exercise will be presented as a new round. Lastly, two status bars will be shown to indicate the progress of both each exercise (round), as well as the total progress of the therapy. Such bars will grow provided the player performed an exercise or a repetition, and their color will turn from white to red as it usually happens in fighting video games.

On the other hand, Influence is the means we use to modify the motivation of the patient towards rehabilitation. Influencing is the power to affect a person or course of events without undertaking any direct action and to be a compelling force on the behavior of others [7]. There are different theories aimed at explaining how someone's motivation can be influenced. Probably one of the most well-known theories regarding influence in motivation is the Theory of Influence [7]. One of the problems of using these motivation theories in the raw is how to put them into practice in a software development. Influence Awareness [19] gathers different motivation theories into a framework. It enumerates different features that can be included to influence user's motivation. It does not prescribe how to implement the influence features but provides a scaffold to properly consider motivation in any software application development.

The design of how to influence the motivation of the patient has been organized around two main axes of Influence Awareness: *authority* and *consistency*. It is important to highlight that the amount of elements used to influence must be carefully selected to avoid (influence) awareness overloading [19]. Authority principles relies on the fact that people usually respect authority. Three elements from this principle were chosen: *ranks*, *experience* and *competence*. Experience is used in the application welcome screen. There, we clearly state that the exercises have been designed by specialist, that is, we stress the idea of a therapy designed by experienced rehabilitation professionals. Ranks makes use of the fact that we tend to obey people with some degree or relevant position. In Fig. 1(a), an image of a person dressed as a doctor is used to explain how the exercise should be performed, since a person wearing a doctor gown and a stethoscope represents the degree and position of the doctor that designed the exercise. Aiming at showing how competent the patient is in doing this exercise, the best previous score achieved by the patient is presented to the user at the top of the screen (see Fig. 1(a)) in the format of the typical high score found in many games. Regarding consistency principle, *creators* elements was chosen, because the chance of the patient rejecting the application concerned us.

Additionally, several influence awareness guidelines [19] were considered during the design of Myo-Rehab. First, the specialists design *challenging tasks*, neither too hard to demotivate patients or too simple to bore the patient. We provide *feedback in the human loop* (see Fig. 1(b)) to keep the patient informed about the progress and achievements during exercise performance. This is achieved by showing the current score (the number of times a movement has been properly done), the progress of the therapy, which is the left red bar, and the progress of the current exercise, represented in the right red bar. Moreover, the user gets haptic feedback from the armband whenever a repetition is made correctly. Lastly, since the patient is watching himself on the screen during exercise performance, he gets live feedback from his/her performance.

4 Conclusions

As was stated in the introduction, giving patients with a proper rehabilitation relay on both considering patients' needs and providing them with proper feedback. However, the repetition of exercises required for achieving motor learning and cortical changes as well as the isolation when patients carry out their therapies at home, led them to give up their rehabilitation.

In this paper we present Myo-Rehab a new tool for designing motivating therapies. It uses Myo, an electromyograph device able to monitor wrist movements that, according to the literature, are required for its rehabilitation. Moreover, in its design motivation has been carefully introduced by means of two different aspects: In Myo-Rehab this issue has been carefully addressed from two points of view: gamification and influence. The former facilitates patients may forget they are involved in a rehabilitation process but enjoying their time with a relaxing activity. The latter is introduced to modify the motivation of the patient towards rehabilitation. As far as we know, this is the first approach combining all these aspects. As future work, we are designing different experiments to evaluate our proposal regarding its usability.

Acknowledgements. This work was partially supported by Spanish Ministerio de Economía, Industria y Competitividad, Agencia Estatal de Investigación (AEI)/European Regional Development Fund (FEDER, UE) under Vi-SMARt (TIN2016-79100-R).

References

1. Aguiar, L.F., Bo, A.P.L.: Hand gestures recognition using electromyography for bilateral upper limb rehabilitation. In: 2017 IEEE Life Sciences Conference (LSC), pp. 63–66. IEEE (2017)
2. Amirabdollahian, F., Walters, M.L.: Application of support vector machines in detecting hand grasp gestures using a commercially off the shelf wireless myoelectric armband. In: 2017 International Conference on Rehabilitation Robotics (ICORR), pp. 111–115 (2017)
3. Batista, T.V.V., Machado, L.S., Valenca, A.M.G.: Surface electromyography for game-based hand motor rehabilitation. In: 2016 XVIII Symposium on Virtual and Augmented Reality (SVR), pp. 140–144. IEEE (2016)
4. Bevilacqua, V., Brunetti, A., Trigiante, G., Trotta, G.F., Fiorentino, M., Manghisi, V., Uva, A.E.: Design and Development of a Forearm Rehabilitation System Based on an Augmented Reality Serious Game. Presented at the (2016)
5. Bütefisch, C., Hummelsheim, H., Denzler, P., Mauritz, K.H.: Repetitive training of isolated movements improves the outcome of motor rehabilitation of the centrally paretic hand. J. Neurol. Sci. **130**(1), 59–68 (1995)
6. Charles, S.K., Krebs, H.I., Volpe, B.T., Lynch, D., Hogan, N.: Wrist rehabilitation following stroke: initial clinical results. In: Proceedings of the 2005 IEEE 9th International Conference on Rehabilitation Robotics, pp. 13–16. IEEE (2005)
7. Cialdini, R.B.: Influence: The Psychology of Persuation. Morrow, New York (1993)
8. Cram, J.R., Steger, J.C.: EMG scanning in the diagnosis of chronic pain. Biofeedback Self Regul. **8**(2), 229–241 (1983)
9. Deterding, S., Sicart, M., Nacke, L., O'Hara, K., Dixon, D.: Gamification using game-design elements in non-gaming contexts. In: 2011 Annual Conference Extended Abstracts on Human Factors in Computing Systems (CHI EA 2011), pp. 24–25. ACM Press, Vancouver (2011)
10. Dromerick, A.W., Edwards, D.F., Hahn, M.: Does the application of constraint-induced movement therapy during acute rehabilitation reduce arm impairment after ischemic stroke? Stroke **31**(12), 2984–2988 (2000)
11. Esfahlani, S.S., Thompson, T., Parsa, A.D., Brown, I., Cirstea, S.: ReHabgame: a non-immersive virtual reality rehabilitation system with applications in neuroscience. Heliyon **4** (2), e00526 (2018)
12. He, S., Yang, C., Wang, M., Cheng, L., Hu, Z.: Hand gesture recognition using MYO armband. Chinese Automation Congress (CAC), 2017, pp. 4850–4855 (2017)
13. Holden, M.K.: Virtual environments for motor rehabilitation: review. CyberPsychology Behav. **8**(3), 187–211 (2005)

14. Horger, M.M.: The reliability of goniometric measurements of active and passive wrist motions. Am. J. Occup. Ther. **44**(4), 342–348 (1990)
15. Kingston, B.: Understanding Joints: A Practical Guide to Their Structure and Function. Nelson Thornes (2000)
16. Langan, J., Subryan, H., Nwogu, I., Cavuoto, L.: Reported use of technology in stroke rehabilitation by physical and occupational therapists. Disabil. Rehabil. Assist. Technol. **13**(7), 1–7 (2017)
17. Leap Motion Inc: Leap Motion. https://www.leapmotion.com/
18. Van der Lee, J.H., Wagenaar, R.C., Lankhorst, G.J., Vogelaar, T.W., Devillé, W.L., Bouter, L.M.: Forced use of the upper extremity in chronic stroke patients: results from a single-blind randomized clinical trial. Stroke **30**(11), 2369–2375 (1999)
19. López-Jaquero, V., Montero, F., Teruel, M.A.: Influence awareness: considering motivation in computer-assisted rehabilitation. J. Ambient Intell. Humaniz. Comput. **10**(6), 2018–2197 (2017)
20. Mendez, I., Hansen, B.W., Grabow, C.M., Smedegaard, E.J.L., Skogberg, N.B., Uth, X.J., Bruhn, A., Geng, B., Kamavuako, E.N.: Evaluation of the Myo armband for the classification of hand motions. In: 2017 International Conference on Rehabilitation Robotics (ICORR), pp. 1211–1214 (2017)
21. World Health Organization: International Classification of Functioning, Disability and Health: ICF. World Health Organization (2001)
22. Ortiz-Catalan, M., Nijenhuis, S., Ambrosch, K., Bovend'Eerdt, T., Koenig, S., Lange, B.: Virtual reality. In: Emerging Therapies in Neurorehabilitation, pp. 249–265. Springer (2014)
23. Rechy-Ramirez, E.J., Marin-Hernandez, A., Rios-Figueroa, H.V.: A human-computer interface for wrist rehabilitation: a pilot study using commercial sensors to detect wrist movements. Vis. Comput., 1–15 (2017)
24. Sathiyanarayanan, M., Rajan, S.: MYO Armband for physiotherapy healthcare: A case study using gesture recognition application. In: 2016 8th International Conference on Communication Systems and Networks (COMSNETS), pp. 1–6 (2016)
25. Skirven, T.M., Osterman, A.L., Fedorczyk, J.M., Amadio, P.C.: Rehabilitation of the Hand and Upper Extremity. Mosby (2011)
26. Slutsky, D.J., Herman, M.: Rehabilitation of distal radius fractures: a biomechanical guide. Hand Clin. **21**(3), 455–468 (2005)
27. Tabor, A., Bateman, S., Scheme, E., Flatla, D.R., Gerling, K.: Designing game-based myoelectric prosthesis training. In: Proceedings of the 2017 CHI Conference on Human Factors in Computing Systems - CHI 2017, pp. 1352–1363. ACM Press, New York (2017)
28. Teruel, M.A., Navarro, E., González, P., López-Jaquero, V., Montero, F.: Applying thematic analysis to define an awareness interpretation for collaborative computer games. Inf. Softw. Technol. **74**, 17–44 (2016)
29. Thalmic Labs Inc.: Myo Gesture Control Armband
30. Vines, A.: Helping your wrist to recover after a fracture. Oxford University Hospitals NHS Trust (2015)
31. Wolf, S.L., Winstein, C.J., Miller, J.P., Taub, E., Uswatte, G., Morris, D., Giuliani, C., Light, K.E., Nichols-Larsen, D.: EXCITE investigators, for the: effect of constraint-induced movement therapy on upper extremity function 3 to 9 months after stroke: the EXCITE randomized clinical trial. J. Am. Med. Assoc. **296**(17), 2095–2104 (2006)
32. Zhou, H., Hu, H.: Human motion tracking for rehabilitation—a survey. Biomed. Signal Process. Control **3**(1), 1–18 (2008)

Motorized Circular Rail with RGB-D Sensor on Cart for Physical Rehabilitation

Ramón Panduro[1], Lidia M. Belmonte[1,2], Eva Segura[2], Paulo Novais[3], José Pascual Molina[1,4], Pascual González[1,4], Antonio Fernández-Caballero[1,2](✉), and Rafael Morales[1,2]

[1] Instituto de Investigación en Informática de Albacete, Universidad de Castilla-La Mancha, 02071 Albacete, Spain
ramon.panduro@alu.uclm.es, {LidiaMaria.Belmonte, josepascual.molina, pascual.gonzalez, Antonio.Fdez, rafael.morales}@uclm.es

[2] Escuela Técnica Superior de Ingenieros Industriales, Universidad de Castilla-La Mancha, 02071 Albacete, Spain
Eva.Segura@uclm.es

[3] Intelligent Systems Lab, Universidade do Minho, Campus of Gualtar, 4710-057 Braga, Portugal
pjon@di.uminho.pt

[4] Escuela Superior de Ingeniería Informática, Universidad de Castilla-La Mancha, 02071 Albacete, Spain

Abstract. This paper introduces a motorized circular rail managing the movement of two carts equipped with an RGB-D sensor each. The proposal is aimed at continuously tracking a person who is undergoing a series of physical rehabilitation exercises from two different viewpoints to monitor if the exercises are being correctly conducted. More concretely, this work offers all details of the trajectory calculation for safe movement of both carts on the motorized circular rail. Then, two study cases are presented to show the efficiency of the control algorithms implemented.

Keywords: Physical rehabilitation · Moving cart · Motorized circular rail

1 Introduction

Assistance and rehabilitation systems based on computers have now become popular. These approaches typically use a depth camera that captures the figure of a human and transmits it to a processing node [1,2]. Precisely, our research team has been working in providing RGB-D solutions for rehabilitation exercises (e.g. [3,4]) in the last years. This kind of vision-based systems are based on human detection [5,6] and tracking [7–9].

Nonetheless, it is mandatory to obtain good viewpoints from the tracked people who are exercizing themselves during physical rehabilitation programs.

P. Novais et al. (Eds.): ISAmI 2019, AISC 1006, pp. 207–215, 2020.
https://doi.org/10.1007/978-3-030-24097-4_25

The RGB-D sensor has to be placed on the most convenient position in an intelligent manner to capture the most relevant body parts of the patient depending on the type of exercise. This is why, we are also developing mechanical solutions for moving RGB-D sensors during human rehabilitation exercises [10,11].

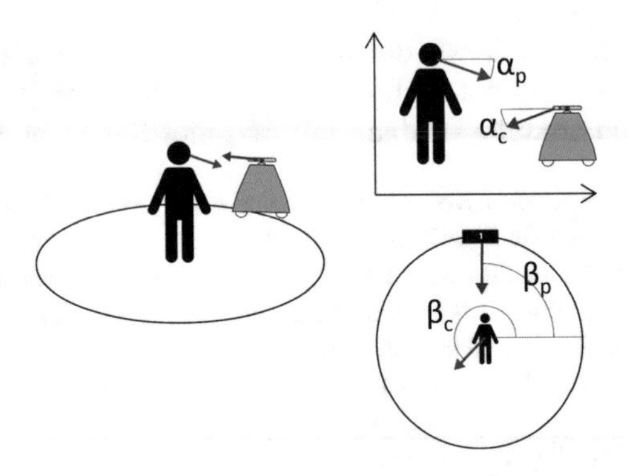

Fig. 1. Vector schemes.

This article describes the operation of a motorized circular rail capable of moving in an intelligent way a couple of carts equipped with an RGB-D camera each for monitoring the patient's body from two complementary views [12]. In this design, the patient is always in the center of the circle and the carts are arranged around him/her so as not to lose sight of the tracking vectors.

2 Trajectory Calculation

The control system of the motorized circular rail must calculate the trajectory of the two carts to place them in the ideal position to track the patient. The patient is introduced into the system as an input vector. It is assumed that both the patient's tracking vectors and the position vectors of the two carts are known in each instant. Several parameters are considered so that the trajectory is carried out correctly: the tilt and pan movements of the cameras (one per cart), the final angular position of the carts with respect to the rail, the vector that each cart must follow, and the possible collisions between both carts.

The formulas that govern the behavior of the carts are different depending on their master or slave role assigned. Thus, it is necessary to establish these roles according to the situation in which both carts are found. On the other hand, the angular position has to vary when a collision with the master cart is foreseen in the calculated trajectory of the slave. In this case, the slave must follow the opposite path.

The movement of the two carts (Q and S) is defined in relation to the tilt and the angular position. Since the patient is always in the center of the circular rail, the pan movement is not necessary and it is enough for the camera to remain

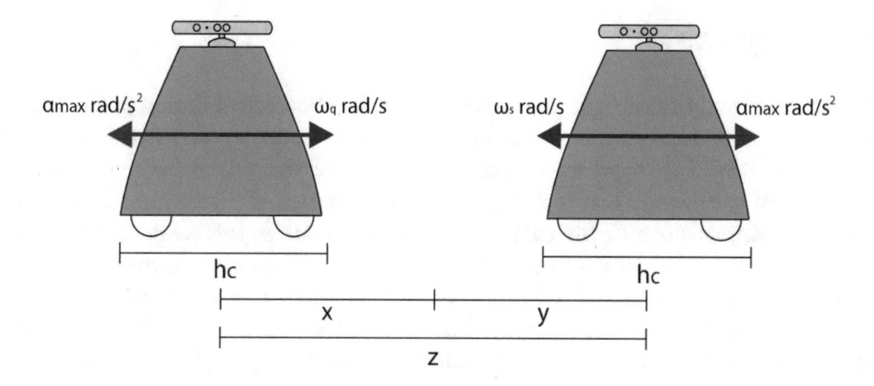

Fig. 2. Security distance of the carts.

straight. Therefore, in first place, the angular position that each cart must take is calculated. Then, the tilt movement that the RGB-D camera must make to focus on the desired patient's body area. Figure 1 shows the nomenclature of angles and vectors that should be considered in the calculations.

In addition, a safety distance is added to ensure that a collision of the slave with the master cart will not occur. The maximum distance at which the two carts would collide in the most extreme case is calculated to obtain a safety zone. This corresponds to bot carts moving at full speed and slowing down with maximum acceleration to reach zero speed. The maximums of speed and angular acceleration are $\alpha_{max} = \pi/4$ rad/s^2 and $\omega_{max} = \pi/2$ rad/s, respectively. The distance reached by a cart until it brakes is called θ.

$$0 = \omega_{max}^2 - 2 \cdot \alpha_{max} \cdot \theta \; ; 0 = (\frac{\pi}{2})^2 + (\frac{-2\pi}{4}) \cdot \theta \; ; \theta = \frac{\pi}{2} \tag{1}$$

The width h_c of the carts is subtracted from the safety distance. Then, it is multiplied by a safety factor λ_s. Obviously, it would be interesting to estimate the safety distance according to the actual speed of both carts. Figure 2 shows the scheme representing the safety distance considering the carts' angular velocities ω_i, where

$$x = \frac{\omega_q^2}{2\alpha_{max}} \; ; y = \frac{\omega_s^2}{2\alpha_{max}} \; ; z = \frac{\omega_q^2 + \omega_s^2}{2\alpha_{max}} = \frac{\omega_q^2 + \omega_s^2}{\pi/2} \tag{2}$$

The angle that covers the width of a cart, ω_c, is defined by the following formula:

$$\omega_c = \frac{h_c}{2R} \tag{3}$$

being $2\omega_c$ the total safety angle for both carts. Subtracting this distance and multiplying by a safety factor λ_s, the final value of safety distance z_s is gotten as

$$z_s = (\frac{\omega_q^2 + \omega_s^2}{2\alpha_{max}} - \frac{h_c}{R}) \cdot \lambda_s \tag{4}$$

3 Study Cases

A couple of study cases are shown to validate the theoretical design of the tracking trajectories. Four different graphs are provided in both study cases. The first graph consists of a two-dimensional aerial view of the initial instant (starting positions of the carts). The second graph also uses an aerial view, but now showing the final positions of the carts after moving to track the patient. The other two graphs are used to check the tilt of the RGB-D camera. Two-dimensional views offer an initial and final graph of the cameras' vectors. A simplified tilt system is considered where a cart does not affect the position of its camera. Moreover, it is considered that the cameras are observing each other when their vectors have the same direction but alternate sense.

Master vs. slave cart role assignment is an important issue. This is why a complete analysis is carried out in both proposed study cases. Study case A shows a normal tracking situation whilst study case B bases on a possible collision situation. The numerical positioning data in both situations are shown in Table 1.

Table 1. Study data.

Data (rad)	Study case A	Study case B
β_{PM}	$3\pi/4$	$3\pi/4$
β_{PE}	0	π
β_{CQ}	$\pi/2$	$\pi/2$
β_{CS}	$\pi/4$	$\pi/4$
α_{PM}	$\pi/4$	$\pi/4$
α_{PE}	$\pi/7$	$\pi/7$
α_{CQ}	$\pi/3$	$\pi/3$
α_{CS}	$-\pi/5$	$-\pi/5$

3.1 Study Case A

A tracking system has been tested with an immobile person in the centre of the rail. Thus, it is always possible to follow the patient's vectors (main and secondary) by keeping the camera pointing to the center. The values shown in Table 1 have been inserted as input values, generating the graphs illustrated in Fig. 3. The first graph shows the start moment, with cart Q depicted as a yellow vector and cart S in blue. The red vector symbolizes the main vector of the patient, while the green one is the secondary. The priority for the system is to choose the most efficient path to cover both vectors.

First of all, the roles that shortens the path taken by both carts as much as possible are defined for Q and S. In case this distance is the same and there is

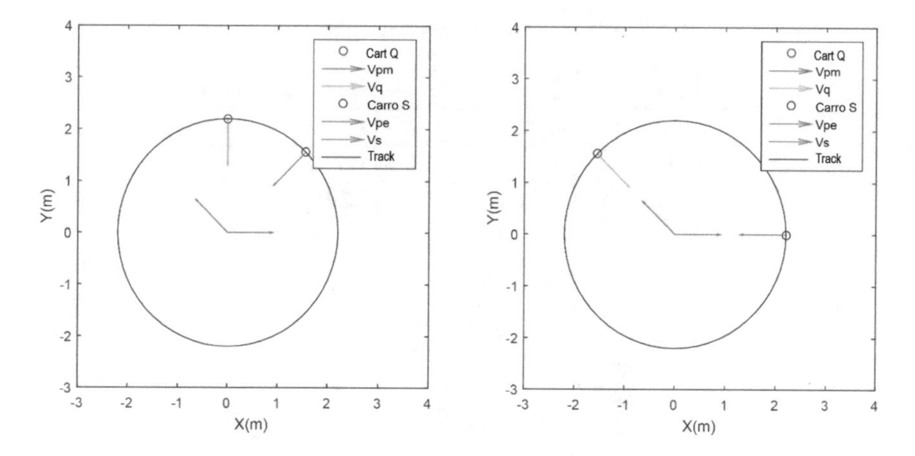

Fig. 3. Positioning graphs of study case A. Left: start positions. Right: final positions.

no possibility of collision, cart Q takes the master role. In this particular study case A, the system is in a normal situation, the shortest path calculation gives the role of master to cart Q and cart S is the slave. As shown in Fig. 3, the roles have been correctly assigned and the carts have moved to the expected final positions. In order to check if the system has selected the shortest path, Table 2 shows a comparison between the values calculated by the system and the selected final path. In both cases, the system selects the shortest path between the two possible paths, depending on whether clockwise or counterclockwise direction is chosen for carts Q and S.

Table 2. Distance according to path.

Data (rad)	Cart Q	Cart S
Path clockwise	−5.4978	−0.7854
Path counterclockwise	0.7854	5.4978
Path selected	0.7854	−0.7854

Secondly, it is interesting to check if the roles are correctly assigned so that the total distance traveled by both carts is minimal (see Table 3). Since cart Q is selected as master and S as slave, the system has correctly selected the shortest distance traveled by both carts.

The next issue to check is the tilt of the camera. After entering the data shown in Table 3, the graphs shown in Fig. 4 are obtained.

The camera has to tilt in order to point directly at the area that is to be observed as smallest path. As can be seen in the graph, the vector of cart Q is tilted until it is in the same direction as the patient's main vector, but in the opposite direction. In the same way, the vector of cart S is tilted to observe the

Table 3. Distance according to role.

Data (rad)	Q master, S slave	Q slave, S master
Distance Q	0.7854	1.5708
Distance S	0.7854	1.5708
Distance total	1.5708	3.1416

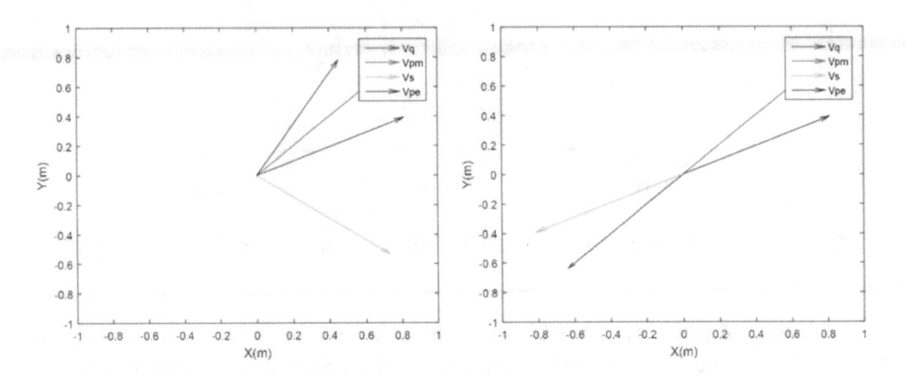

Fig. 4. Tilt graphs of study case A. Left: start positions. Right: final positions.

patient's secondary vector. As with positioning, the system is able to select the shortest path (see Table 4). In this case, it is counterclockwise and clockwise for cart Q and cart S, respectively.

Table 4. Tilt according to path.

Data (rad)	Cart Q	Cart S
Path clockwise	−3.4034	−2.0645
Path counterclockwise	2.8798	4.2187
Path selected	2.8798	−2.0645

The final displacement functions of carts Q and S are called M_{CQ} and M_{CS}, respectively.

$$M_{CQ}(\beta, \alpha) = (0.7854, 2.8798) \text{ rad} \; ; \; M_{CS}(\beta, \alpha) = (-0.7854, -2.0654) \text{ rad} \quad (5)$$

3.2 Study Case B

In this new study case, the position of β_{PE} is modified to cause a collision possibility and thus to observe the system's reaction. It is checked how the change of roles affects the positioning. The graphs shown in Fig. 5 allow observing the evolution of the system with the new position of the patient's secondary vector.

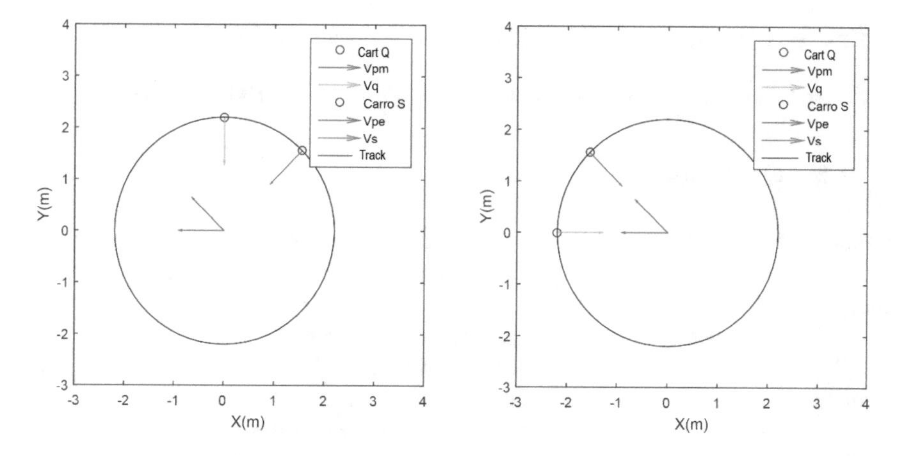

Fig. 5. Positioning graphs of study case B. Left: start positions. Right: final positions.

Here, the roles are changed as consequence of the anticipation of a collision based on the calculation of the shortest distance. Collision means that one of the carts will interfere in the trajectory of the other. In this case, cart Q is placed as slave and S as master. This arrangement is initiated by introducing the camera data as input vectors. However, these data have been considered as the position of the carts to simplify the selection of roles. Next, the tilt of the cameras are evaluated as shown in Fig. 6.

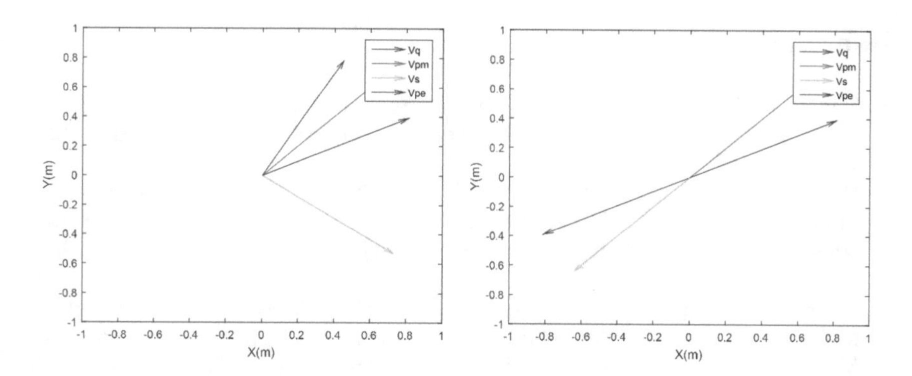

Fig. 6. Tilt graphs of study case B. Left: start positions. Right: final positions.

As explained above, the choice of the tracking vectors is defined by the angular positions of the carts. In this case, the tilt has a different path than in study case A, even as they have to go to the same place as before. Here, it is cart Q that follows the patient's secondary vector, while the S cart tracks the patient's main vector. It can be seen in the graphs that the tracking task is done correctly.

In this case, the movement functions of the carts are as follows:

$$M_{CQ}(\beta, \alpha) = (1.5708, 2.5432) \text{ rad} \; ; M_{CS}(\beta, \alpha) = (1.5708, -1.7279) \text{ rad} \quad (6)$$

4 Conclusions

This paper has introduced a trajectory calculation algorithm for the movement of two carts (equipped with an RGB-D sensor) on a motorized circular rail. The final application of such couple of carts is continuous monitoring of a person who is fulfilling a series of physical rehabilitation exercises.

The paper has also described two study cases that show that the proposed algorithm possesses enough security to ensure that both carts will never collide during their tracking of the rehabilitation program of the human.

As future work, we propose: (a) to extend the trajectory planner to consider the free movement of the person within the working area of the proposed rail system; (b) to study different sensory strategies for conducting experimental trials that allow patients to test the system; (c) to conduct studies on users' perception for this kind of monitoring systems in order to draw conclusions about possible design improvements.

Acknowledgements. This work was partially supported by Ministerio de Ciencia, Innovación y Universidades, Agencia Estatal de Investigación (AEI)/European Regional Development Fund (FEDER, UE) under DPI2016-80894-R and TIN2016-79100-R grants.

References

1. Chang, Y., Chen, S., Huang, J.: A Kinect-based system for physical rehabilitation: a pilot study for young adults with motor disabilities. Res. Dev. Disabil. **32**(6), 2566–2570 (2011)
2. Freitas, D., Gama, A. Da, Figueiredo, L., Chaves, T., Marques-Oliveira, D., Teichrieb, V., Araújo, C.: Development and evaluation of a Kinect based motor rehabilitation game. Simposio Brasileiro de Jogos e Entretenimento Digital, pp. 144–153 (2012)
3. Oliver, M., Montero, F., Molina, J.P., González, P., Fernández-Caballero, A.: Multi-camera systems for rehabilitation therapies: a study of the precision of Microsoft Kinect sensors. Front. Inform. Technol. Electron. Eng. **17**(4), 348–364 (2016)
4. Oliver, M., Montero, F., Fernández-Caballero, A., González, P., Molina, J.P.: RGB-D assistive technologies for acquired brain injury: description and assessment of user experience. Expert Syst. **32**(3), 370–380 (2015)
5. Fernández-Caballero, A., Castillo, J.C., Serrano-Cuerda, J., Maldonado-Bascón, S.: Real-time human segmentation in infrared videos. Expert Syst. Appl. **38**(3), 2577–2584 (2011)
6. Fernández-Caballero, A., López, M.T., Saiz-Valverde, S.: Dynamic stereoscopic selective visual attention (DSSVA): integrating motion and shape with depth in video segmentation. Expert Syst. Appl. **34**(2), 1394–1402 (2008)

7. Castillo, J.C., Fernández-Caballero, A., Serrano-Cuerda, J., López, M.T., Martínez-Rodrigo, A.: Smart environment architecture for robust people detection by infrared and visible video fusion. J. Ambient Intell. Humanized Comput. **8**(2), 223–237 (2017)
8. Fernández-Caballero, A., López, M.T., Serrano-Cuerda, J.: Thermal-infrared pedestrian ROI extraction through thermal and motion information fusion. Sensors **14**(4), 6666–6676 (2014)
9. Moreno-Garcia, J., Rodriguez-Benitez, L., Fernández-Caballero, A., López, M.T.: Video sequence motion tracking by fuzzification techniques. Appl. Soft Comput. **10**(1), 318–331 (2010)
10. Panduro, R., Oliver, M., Morales, R., González, P., Fernández-Caballero, A.: Motorized multi-camera slider for precise monitoring of physical rehabilitation. Lecture Notes in Computer Systems, vol. 10070, pp. 21–27 (2016)
11. Gascueña, J.M., Fernández-Caballero, A.: Agent-oriented modeling and development of a person-following mobile robot. Expert Syst. Appl. **38**(4), 4280–4290 (2011)
12. Mkhitaryan, A., Burschka, D.: RGB-D sensor data correction and enhancement by introduction of an additional RGB view. In: Proceedings of the IEEE/RSJ International Conference on Intelligent Robots and Systems, pp. 1077–1083 (2013)

Assisting Dependent People at Home Through Autonomous Unmanned Aerial Vehicles

Lidia M. Belmonte[1,2], Rafael Morales[1,2], Arturo S. García[1,2], Eva Segura[1,2], Paulo Novais[3], and Antonio Fernández-Caballero[1,2,4](✉)

[1] Instituto de Investigación en Informática de Albacete, Universidad de Castilla-La Mancha, 02071 Albacete, Spain
{rafael.morales,arturosimon.garcia,eva.segura,Antonio.Fdez}@uclm.es

[2] Escuela Técnica Superior de Ingenieros Industriales, Universidad de Castilla-La Mancha, 02071 Albacete, Spain
lidiamaria.belmonte@uclm.es

[3] Escola de Engenharia, Universidade do Minho, Campus de Gualtar, 4710-057 Braga, Portugal
pjon@di.uminho.pt

[4] CIBERSAM (Biomedical Research Networking Centre in Mental Health), Madrid, Spain

Abstract. This work describes a proposal of autonomous unmanned aerial vehicles (AUAVs) for home assistance of dependent people. AUAVs will monitor and recognize human activities during flight to improve their quality of life. However, before bringing such AUAV assistance to real homes, several challenges must be faced to make them viable and practical. Some challenges are technical and some others are related to human factors. In particular, several technical aspects are described for AUAV assistance: (1) flight control, based on our active disturbance rejection control algorithm, (2) flight planning (navigation in obstacle environments), and, (3) processing signals, acquired both from flight-control and monitoring sensors. From the assisted person's viewpoint, our research focuses on three cues: (1) the user's perception about AUAV assistance, (2) the influence on human acceptance of AUAV appearance and behavior at home, and (3) the human-robot interaction between assistant AUAV and assisted person. Finally, virtual reality environments are proposed to carry out preliminary tests and user acceptance evaluations.

Keywords: Autonomous unmanned aerial vehicles · Home assistance · Dependent people

1 Introduction

The use of unmanned aerial vehicles (UAVs) has notably increased in the last years. Moreover, computer vision in UAVs plays a role beyond serving as mere recording and displaying of flight environments. By means of computer vision algorithms,

P. Novais et al. (Eds.): ISAmI 2019, AISC 1006, pp. 216–223, 2020.
https://doi.org/10.1007/978-3-030-24097-4_26

it is possible to extract useful information both of the aircraft's state and of its environment. This article proposes a framework for assisting dependent people at home through vision-based autonomous unmanned aerial vehicles (AUAVs) which do not require the presence of an operator and navigate indoor without contravening current laws for flying outdoors. Thus, the overall objective is to enhance dependent people's quality of life (QoL). QoL is the appreciation of well-being in daily human lives, including emotional, social and physical aspects. QoL of dependent people is usually reduced as a consequence of their functional incapacity for carrying normal daily activities. In addition, dependents often prefer to live in their own homes against other options, but it is difficult to provide the necessary security and home care without monitoring [1,2]. In this sense, the combination of information and communication technologies with mobile robotics provides intelligent and proactive actions to most problems that dependent people suffer at home. Thus, they facilitate the care of dependent people so that they improve their QoL in the comfort of their proper homes.

It is our conviction that aerial vehicles can complement other technologies in assisting dependents at home. Indeed, AUAVs act alone or in combination with other technologies like surveillance cameras and biometric bracelets. They also reach out where other solutions do not (for example, blind zones) through accompanying the person, positioning itself to observe his/her activities, evaluating his/her emotional state, and acting in accordance to each situation. For this sake, the main challenges for AUAVs to assist dependents within a family environment are described throughout this article. Explicitly, a double engineering solution to put in practice the above mentioned objective is described.

In first place, we have detected important technical challenges in terms of flight control, flight planning in environments with obstacles and signal processing. The solution for robust flight at home is active control by disturbance rejection, our proper algorithm recently introduced [3]. Signal processing is required for sensors that control the flight itself (e.g. inertial measurement unit sensors) and also those that observe the habitat. Obviously, the observed environment includes the dependent so that an on-board camera will capture the person's activities and facial expressions.

Secondly, human factors must consider the individual not just as an obstacle, but respecting his/her personal space when calculating the flight paths. In addition, the potential prejudices and doubts that a person has towards a flying robot, probably considered as an intruder or threat, have to be saved. Lastly, the most appropriate human-robot interaction between assisted and assistant must be defined to build a relationship of trust.

Finally, such proposal has to be evaluated. In this initial stage, the focus is put towards the acceptance of AUAVs as assistants, deepening in the user's response in immersive virtual reality environments to AUAV appearance and behavior, as well as the interaction between human and robot.

2 Technical Challenges

The development of AUAVs to serve as assistants for dependents at their own homes requires the design of a solution for autonomous and safe flight of the aircraft. The flight environment, that is the dependent person's home, is initially a place containing static obstacles from the viewpoint of navigation. However, the dependent person also moves around the environment as a dynamic obstacle that must be followed by the AUAV. In this context, the use of multi-rotor AUAVs has been considered appropriate due to their excellent maneuverability, agility, and versatility.

In addition, it is mandatory to reduce the AUAV's size in order not to interfere with people's daily routines in limited home space. However, such reduction makes difficult its control. Mainly, the sensitivity of an AUAV is highly affected both by exogenous (wind gusts) and endogenous (large non-linearities, uncertainties, dynamic couplings, etc.) disturbances that seriously affect its flight capacity and stability [4]. Several works have been developed to reduce the effects of wind gusts on the AUAV [13], but most of them are based on the assumption of persistent gusts of wind with a fixed speed, a fact that almost never occurs in reality. There are also enormous limitations on the available space, payload, and capacity of an AUAV's power supply system, leading to the use of small processors with low power consumption and limited memory [5].

On the other hand, the integration of AUAVs within the dependent person's home must be based on the principle that safety is not compromised during flight, exhibiting a level of safety equivalent to that of manned flight missions [7,8]. All this makes it necessary to design robust control algorithms that can be implemented in real small-size AUAVs [9]. Unfortunately, conventional control methods (*proportional integral derivative* (PID) and *linear quadratic* (LQR)) present serious performance problems when the size of AUAVs is reduced [10]. Therefore several methods have been developed to improve the control performance in multi-rotor AUAVs. These are non-linear type controllers based on mathematical models obtained through using complicated non-linear models and identification methods [11]. Many of them present highly complex problems, which complicates their use in real AUAVs with low consumption processors and limited memory. Another disadvantage is that the multi-rotor AUAV's dynamic model is an approximation to the real system, showing therefore problems due to parametric uncertainties and noise in sensors' measures.

In addition, robust algorithms of an adaptive nature have been developed in recent years to address the problem of parametric uncertainties in AUAVs. Unfortunately, they present problems when the multi-rotor AUAV navigates in outdoor environments (including indoors with open windows), as these control systems are affected by external wind disturbances [12]. In this sense, the development of controllers for multi-rotor AUAVs performing aggressive maneuvers under disturbances due to variable speed wind is now a completely open field.

In addition, the measurements from sensors have a high noise component, especially when the sensors are low cost [6]. In fact, signals can cause the control system to become unstable when used in feedback or compensation loops. The search for methods to eliminate noise from signals through hardware and/or software is a field of current research in many application domains. In this way, a prevailing research field is to obtain signal filters that hardly present delays, have high robustness with respect to noise, do not need to presuppose statistical properties of noise, and can be implemented in low-cost hardware systems online and in real-time. Thus, a first major challenge for the development of our proposal is the design of innovative robust flight control and signal processing systems to be implemented on small processors with reduced energy consumption and limited memory.

We propose to improve the efficiency and robustness of the AUAV against significant uncertainties in its modeling and external disturbances to solve this first challenge. This has been done through the development of new algorithms based on the concept of active disturbance rejection control (ADRC) [3] and its experimentation with the Twin Rotor MIMO System (TRMS) (see Fig. 1).

Fig. 1. TRMS system used in flight control simulations

ADRC handles the effects of disturbances (endogenous and exogenous) as an aggregate, global, purely time-dependent function without a particular structure. This concept allows the designer to avoid the use of an observer based on the non-linear structure of the (often non-existent) system and proposes a non-linear injection module of the inputs through their gain factors, instead of a merely linear observer. The arbitrarily close estimate of the aggregate disturbance allows its approximate cancellation through the appropriate control action. With this new concept it is intended to attain: (a) an improvement in the efficiency of the AUAV behavior, (b) a higher tolerance to large disturbances in the AUAV, (c) an easier adjustment and operation, and, (d) a drastic reduction of the AUAV probability to go into loss when performing home monitoring tasks.

3 Human Factor Challenges

Automated monitoring and identification of humans is a valuable tool in many areas such as rehabilitation, clinical psychology and gerontology for taking care in the family environment [14]. The availability of new static and mobile sensor types, and the consequent fusion of multi-sensory data in monitoring tasks offers novel solutions to model environments and diagnose situations based on the analysis of sensory data sequences [15].

Monitoring people physical activities and behaviors from computer vision is now well-established in our team [16]. For recognizing the emotional state of the human, a non-intrusive process is automatic detection of emotions based on the study of facial expressions [17]. The Facial Action Coding System encodes all possible facial expressions according to action units (AUs) that occur individually or in combination [18]. The exploitation of these context-aware emotional devices allows to deliver a highly personalized and dedicated collection of services designed to support users and improve their personal care.

Thus, the AUAV performs a capture of images of the dependent person's face. This requires the detection and focus on the principal parts of the face (eyes, nose, mouth, etc.). All this requires the positioning of the AUAV in front of the person's face from time to time. The captured information is sent to a base station responsible for the recognition of the dependent's emotions. In this way, the system evaluates the person's emotional state, together with his/her behavior, to determine the assistance needed for each situation.

Let us highlight that the ability of an UAV to fly autonomously is essential to carry out the proposed monitoring tasks. This ability is mandatory when considering that homes are closed environments composed of static and dynamic obstacles of different sizes and types, which endangers the safety of the monitored people. In the scientific literature, there are two approaches to autonomous navigation. (a) In deliberative motion planning the trajectories are obtained assuming a global knowledge of the environment, that is, a static environment [19]. In general, a deliberate trajectory planner is useful when the environment is known a priori, but may require too much computational effort when the environment is dynamic. (b) Reactive motion planning considers obstacle detection by means of a local sensory system (laser, optical flow sensors, stereo cameras or a single camera) and the consequent control of the trajectory to avoid the detected obstacles [20]. In this case, the information is incomplete and uncertain, and suffers from the additional problem of the difficult specification of direct movement plans.

In this sense, we propose to improve the navigation of the AUAV in environments with static and dynamic obstacles by developing a trajectory planner that takes into account the AUAV's dynamic and energetic constraints, and that contemplates the appearance of unexpected obstacles through computer vision and its corresponding integration in the AUAV control algorithm. In this respect, a navigation system composed of different stages is proposed. In the first stage it is assumed that the physical navigation environment is known (map of the home). Several methodologies for obtaining trajectories will be studied by means of optimization algorithms that result in AUAV trajectories free of obstacles. During the second stage new algorithms will be developed to provide the AUAV with the ability to adapt to dynamic environments by integrating the information coming from the sensory system to monitor the dependent person at home.

An interesting example on trajectory planning has been presented for autonomously monitoring wind turbine blades by means of a quadrotor following a hyperbolic path around the blade [21]. UAVs for monitoring large and inaccessible structures have been also used by health, safety and environment inspectors in complex constructions such as power stations [22]. Recently, a novel framework has been developed to increment the ability of autonomous navigation, especially in cluttered environments [23]. Hence, robust and fast motion close to obstacles has been achieved, which demonstrates that a deeper integration of motion planning and perception improves robustness and computational efficiency. Future flight safety improvements will allow to extent the use of AUAVs in complex environments such as cities.

4 Virtual Reality Validation Process

The research work in AUAV assistance requires the validation of the technology and the acceptance of the dependent people. The obtained conclusions will allow to improve further developments. The validation should be carried out at homes of people willing to participate. Nevertheless, in this initial stage of development, it has been considered more appropriate to use simulations. This option saves costs, is more versatile and, most important, is much safer.

For this, we rely on virtual reality (VR) to perform evaluations, focused both on the flying robot assistant and on the assisted person. Thus, virtual worlds will be generated to process the sensor signals and evaluate the progress in flight control and planning. Virtual environments will recreate the dependent and his/her home (a closed environment with obstacles). The behavior of the AUAV will be simulated in different scenarios (see Fig. 2) using MATLAB and Unity 3D.

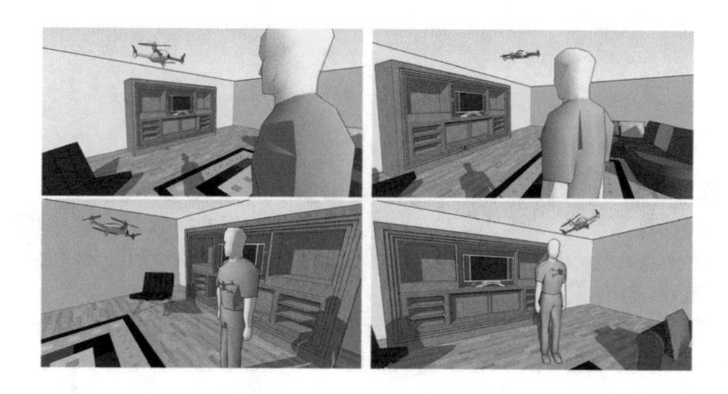

Fig. 2. VR environment to simulate AUAV assistance

Regarding a user validation of the proposal, the focus should be towards the acceptance of AUAVs as assistants, deepening in the users' responses to AUAV appearance and behavior, and human-robot interaction. Through the inclusion of immersive VR headsets and semi-immersive VR technologies, it is possible for real people to experience first hand sharing their space with AUAVs assistants. Even more, VR allow to observe the reaction of people to the flight of different virtual AUAVs around them and to investigate on several human-robot interaction techniques.

5 Conclusions

Personal dependency is defined as a functional incapacity for the development of daily-life activities, which requires assistance for their realization. However, dependent people usually prefer to live at their own homes, which implies care strategies in the family environment. This paper has proposed a solution for the development of indoor AUAV-based systems that allow home assistance of dependents with the aim of improving their quality of life.

Concretely, the use of AUAVs based on computer vision is proposed for the support and help of dependents. This proposal aims to develop UAVs capable of flying autonomously in a home to perform the task of monitoring and assisting dependents. However, this development process entails multiple challenges, both technical and human, that have to be addressed before making possible the use of AUAVs as assistants at home (e.g. a safety radius must be considered during the whole monitoring process to avoid collisions between UAV and person). In this article, we have introduced the bases for the development of solutions to advance in the line of research proposed. We have considered the aspects of AUAV control and navigation, as well as human monitoring. Lastly, virtual reality has been proposed as a key element in the validation of the robotic assistance system.

Acknowledgments. This work has been partially supported by Spanish Ministerio de Ciencia, Innovación y Universidades, Agencia Estatal de Investigación (AEI)/European Regional Development Fund (FEDER, UE) under DPI2016-80894-R grant, and by CIBERSAM of the Instituto de Salud Carlos III. Lidia M. Belmonte holds FPU014/05283 scholarship from Spanish Ministerio de Educación y Formación Profesional.

References

1. Fernández-Caballero, A., Martínez-Rodrigo, A., Pastor, J.M., Castillo, J.C., Lozano-Monasor, E., López, M.T., Zangróniz, R., Latorre, J.M., Fernández-Sotos, A.: Smart environment architecture for emotion recognition and regulation. J. Biomed. Inform. **64**, 55–73 (2016)
2. Castillo, J.C., Castro-González, Á., Fernández-Caballero, A., Latorre, J.M., Pastor, J.M., Fernández-Sotos, A., Salichs, M.A.: Software architecture for smart emotion recognition and regulation of the ageing adult. Cogn. Comput. **8**(2), 357–367 (2016)
3. Belmonte, L.M., Morales, R., Fernández-Caballero, A., Somolinos, J.A.: A tandem active disturbance rejection control for a laboratory helicopter with variable speed rotors. IEEE Trans. Ind. Electron. **63**(10), 6395–6406 (2016)
4. Mahony, R., Kumar, V., Corke, P.: Multirotor aerial vehicles: modeling, estimation, and control of quadrotor. IEEE Robot. Autom. Mag. **19**(3), 20–32 (2012)
5. Leishman, R.C., MacDonald, J.C., Beard, R.W., McLain, T.W.: Quadrotors and accelerometers: state estimation with an improved dynamic model. IEEE Control Syst. Mag. **34**(1), 28–41 (2014)
6. Tanveer, M.H., Ahmed, S.F., Hazry, D., Warsy, F.A., Joyo, M.K.: Stabilized controller design for attitude and altitude controlling of quad-rotor under disturbance and noisy conditions. Am. J. Appl. Sci. **10**(8), 819–831 (2013)

7. Belmonte, L.M., Morales, R., Fernández-Caballero, A., Somolinos, J.A.: Robust decentralized nonlinear control for a twin rotor MIMO system. Sensors **16**(8), 1160 (2016)
8. Yu, Y., Lu, G., Sun, C., Liu, H.: Robust backstepping decentralized tracking control for a 3-DOF helicopter. Nonlinear Dynam. **82**(1–2), 947–960 (2015)
9. Fernández-Caballero, A., Belmonte, L.M., Morales, R., Somolinos, J.A.: Generalized proportional integral control for an unmanned quadrotor system. Int. J. Adv. Robot. Syst. **12**, 85 (2015)
10. Bouabdallah, S., Noth, A., Siegwart, R.: PID vs LQ control techniques applied to an indoor micro quadrotor. In: Proceedings of the 2004 IEEE/RSJ International Conference on Intelligent Robotic Systems, Senday, Japan, pp. 2451–2456 (2004)
11. Bertrand, S., Guenard, N., Hamel, T., Piet-Lahanier, H., Eck, L.: A hierarchical controller for miniature VTOL UAVs: design and stability analysis using singular perturbation theory. Cont. Eng. Pract. **19**(10), 1099–1108 (2011)
12. Dydek, Z.T., Annaswamy, A.M., Lavretsky, E.: Adaptive control of quadrotor UAVs: a design trade study with flight evaluations. IEEE Trans. Cont. Syst. Tech. **21**(4), 1400–1406 (2013)
13. Sun, L., Zuo, Z.: Nonlinear adaptive trajectory tracking control for a quad-rotor with parametric uncertainty. Proc. Inst. Mech. Eng. G **229**(9), 1–13 (2014)
14. Castillo, J.C., Castro-González, Á., Alonso-Martín, F., Fernández-Caballero, A., Salichs, M.A.: Emotion detection and regulation from personal assistant robot in smart environment. In: Costa, A., Julián, V., Novais, P. (eds.) Personal Assistants: Emerging Computational Technologies, pp. 179–195. Springer, New York (2018)
15. Morales, R., Fernández-Caballero, A., Somolinos, J.A., Sira-Ramírez, H.: Integration of sensors in control and automation systems. J. Sensors **2017**, 6415876 (2017)
16. Castillo, J.C., Fernández-Caballero, A., Serrano-Cuerda, J., López, M.T., Martínez-Rodrigo, A.: Smart environment architecture for robust people detection by infrared and visible video fusion. J. Ambient. Intell. Humaniz. Comput. **8**(2), 223–237 (2017)
17. Lozano-Monasor, E., López, M.T., Vigo-Bustos, F., Fernández-Caballero, A.: Facial expression recognition in ageing adults: from lab to ambient assisted living. J. Ambient. Intell. Humaniz. Comput. **8**(4), 567–578 (2017)
18. Ekman, P., Friesen, W.V., Hager, J.C.: Facial Action Coding System. Manual and Investigator's Guide. Research Nexus, Salt Lake City (2002)
19. Lamiraux, F., Sekhavat, S., Laumond, J.P.: Motion planning and control for Hilare pulling a trailer. IEEE Trans. Robot. Autom. **15**(4), 640–652 (1999)
20. Balch, T., Arkin, R.C.: Behavior-based information control for multirobot teams. IEEE Trans. Robot. Autom **14**(6), 926–939 (1998)
21. Shivaram, S.: Structural Health Monitoring of Wind Turbine Blades using Unmanned Air Vehicles. Master's Dissertation, University of Dublin (2015)
22. Whitemore, H.: Koweit: how a drone is being used to monitor Health & Safety at the construction site, ENGIE Innovation (2015). https://innovation.engie.com/en/news/news/smart-buildings/koweit-how-a-drone-is-being-used-to-monitor-health-safety-at-the-construction-site-1/1112
23. Florence, P.R., Carter, J., Ware, J., Tedrake, R.: NanoMap: fast, uncertainty-aware proximity queries with lazy search over local 3D data. In: International Conference on Robotics and Automation (ICRA), Brisbane, Australia (2018)

Author Index

P. Novais et al. (Eds.): ISAmI 2019, AISC 1006, pp. 225–226, 2020.
https://doi.org/10.1007/978-3-030-24097-4

Printed in the United States
By Bookmasters